LAND MINES IN CAMBODIA

The Coward's War

September 1991

Asia Watch
Physicians for Human Rights

Printed in the United States of America

Cover Design By:

ASIA WATCH

Asia Watch, a division of Human Rights Watch, was established in 1985 to monitor and promote in Asia observance of internationally recognized human rights. The Chair is Jack Greenberg and the Vice-chairs are Harriet Rabb and Orville Schell. Sidney Jones is Executive Director. Mike Jendrzejczyk is Washington Director. Patricia Gossman, Robin Munro and Ji Won Park are Research Associates. Julie Triedman is an Orville Schell Fellow. Jeannine Guthrie, Lydia Lobenthal and Emily Sewell are Associates.

PHYSICIANS FOR HUMAN RIGHTS

Physicians for Human Rights is a national organization of health care professionals whose goal is to bring the skills of the medical profession to the protection of human rights. Since its founding in 1986, it has conducted over thirty-five missions concerning twenty-three countries. Physicians for Human Rights adheres to a policy of strict impartiality and is concerned with the medical consequences of human rights abuses regardless of the ideology of the offending government or group. The President of the Board of Directors of Physicians for Human Rights is H. Jack Geiger, M.D.; the Vice President is Carola Eisenberg, M.D. Jonathan Fine, M.D. is Executive Director; Susannah Sirkin is Associate Director; Nancy Arnison is Program Officer; Barbara Ayotte is Publications Assistant.

Library of Congress Cataloging-in-Publication Data

Land mines in Cambodia: the coward's war, September 1991 / Asia Watch, Physicians For Human Rights.
p. cm.
ISBN 1-56432-001-4 :
1. Political atrocities--Cambodia. 2. Cambodia--History--1975- 3.Mines, Military.
I. Asia Watch Committee (U.S.) II. Physicians For Human Rights (U.S.)
DS554.8.L36 1991 91-38936
959.604--dc20 CIP

TABLE OF CONTENTS

GLOSSARY OF ACRONYMS

AFSC	American Friends Service Committee
ANS	Sihanouk National Army (English)
ASEAN	Association of Southeast Asian Nations
CGDK	Coalition Government of Democratic Kampuchea
COERR	Catholic Office for Emergency Relief and Refugees
FAO	Food and Agriculture Organization
FUNCINPEC	United Front for an Independent, Neutral, Peaceful, and Cooperative Cambodia (English)
HI	Handicap International
IED	Improvised Explosive Device
ICRC	International Committee of the Red Cross
IMR	Infant Mortality Rate
JRS	Jesuit Relief Service
KPNLF	Khmer People's National Liberation Front
NCR	Noncommunist Resistance

SAS	Special Air Service
SNC	Supreme National Council
UNBRO	United Nations Border Relief Operation
UNDP	United Nations Development Program
UNHCR	United High Commissioner for Refugees
UNICEF	United Nations Children's Fund
UNTAC	United Nations Transition Authority in Cambodia
WFP	World Food Program

ACKNOWLEDGEMENTS

This report was written by Eric Stover, a freelance writer and consultant to Human Rights Watch and Physicians for Human Rights, and Rae McGrath, director of the Mines Advisory Group, based on research undertaken during missions to Cambodia and Thailand in March and April 1991. Another participant on the mission was Dr. James C. Cobey, a Red Cross consultant and an orthopaedic surgeon from Washington, D.C.

We wish to thank several colleagues who helped in the preparation of this report. Sidney Jones and Michael Jendrzejczyk of Asia Watch; Jemera Rone of Human Rights Watch; and Dr. Jonathan Fine and Richard P. Claude of Physicians for Human Rights reviewed earlier drafts of the manuscript. Nicki Hengen edited the report. Pamela Blotner drew the illustrations of landmines and other devices. John Rogge at the Disaster Research Unit, University of Manitoba, and the Intertect Institute provided us with maps and graphs on the demographics of the Cambodian refugee camps in Thailand. Roger Normand conducted background research in the Thai border camps in March 1991. Asia Watch interns Christina Misa Wong and Marie Tien gathered information on the Cambodian peace process. Asia Watch associate Jeannine Guthrie prepared the manuscript for publication.

We are grateful for the assistance provided to us in Cambodia and Thailand by Médecins du Monde, the International Committee of the Red Cross, and Handicap International.

I. INTRODUCTION

Few countries exemplify mankind's capacity to inflict cruelty upon itself more than Cambodia.[1] In the last 20 years, Cambodia's people have suffered a seemingly relentless series of man-made disasters--from the massive aerial bombardment by the United States in the early 1970s, to the rule of the murderous Khmer Rouge between 1975 and 1979, to the widespread famine which followed, and finally to a savage civil war. Even today, as a cease-fire takes effect and a peaceful settlement to the Cambodian conflict appears likely, government and resistance soldiers continue to lay land mines along footpaths, rice paddies, riverbeds, and around villages. They refer to land mines as their "eternal sentinels," never sleeping, always ready to attack. They see mines as a way of avoiding direct contact with the enemy and so have saturated contested areas with them on a scale unrelated to the actual military need or objective. Unless the mines are cleared and destroyed, they will kill or maim Cambodians well into the next century.

The need for mine clearance and eradication in Cambodia is more urgent now than ever before. In recent months, the United Nations, in anticipation of a peaceful settlement to the war, has begun drafting plans to repatriate hundreds of thousands of Cambodian refugees now living in border camps in Thailand. Many, if not most, of these refugees will return to their farms in what were once combat zones. The chances that they and their families will fall victim to land mines is high.

---

[1] Throughout this report the name **Cambodia** is used rather than **Kampuchea**. In 1989, the government in Phnom Penh formally changed the name of the country from the People's Republic of Kampuchea to the State of Cambodia.

Cambodia, with a population of eight and a half million[2], already has the highest percentage of physically disabled inhabitants of any country in the world--higher even than Afghanistan. There are now over 30,000 amputees in Cambodia, with an additional 5-6,000 living in refugee camps along the Thai border. In 1990 alone, at least 6,000 Cambodians suffered amputations as a result of a mine injury.[3] Most of the casualties were civilians--peasants who stepped on mines while gathering firewood, harvesting rice, herding animals, or fishing. They were the lucky ones; nearly as many died from loss of blood, undiscovered in the fields, or succumbed to their wounds because no transport was available to get them to medical help.

These grim statistics mean that the Cambodian conflict may be the first war in history in which land mines have claimed more victims--combatants and noncombatants alike--than any other weapon.

If the international community waits until a peace settlement is in place before it begins a mines surveying and eradication program, there will be a disaster. Refugees spontaneously returning will fall victim to mines, the clearance program in designated resettlement areas may be too hastily and cheaply put together, and the health care system will be unable to deal with the injuries.

---

[2]There has not been a national census since 1962, at which time the population was 5,728,771. See J. R. Rogge, "Return to Cambodia: The Significance and Implications of Past, Present and Future Spontaneous Repatriations," Disaster Research Unit, University of Manitoba, March 1990, p. 18.

[3]By comparison, in 1989, surgeons in the United States, with a population of 220 million, performed no more than 10,000 amputations on patients who had suffered traumatic injuries.

Several measures should be taken immediately. The five permanent members of the Security Council -- the United States, the Soviet Union, China, Great Britain and France -- should press all warring parties to make a formal commitment, separate from the cease-fire, to stop laying new mines. All outside powers which have supplied mines should provide technical information about the weapons which can help in the clearance process. Plans should be drawn up for an exhaustive province-by-province mines survey, carried out by expert staff with adequate funding. Surveying in some provinces could begin as soon as the field staff are in place. To wait for final agreement to a peace settlement is to wait for more women collecting firewood or more children straying off paths to be blown up.

This report is based on a visit to Cambodia and Thailand in March and April 1991 by a delegation assembled by Asia Watch and Physicians for Human Rights. The purpose was to investigate the use of land mines by all sides in the conflict and their effects on the civilian population. The three members of the delegation were Rae McGrath, director of the Mines Advisory Group; Eric Stover, a writer specializing in medicine and human rights; and James C. Cobey, a Red Cross consultant and an orthopaedic surgeon from Washington, D.C.

In Cambodia, we visited hospitals and prosthetic workshops in the capital of Phnom Penh and in the provinces of Takeo, Battambang, and Banteay Meanchey. Using a set of predetermined questions, we interviewed patients recently injured by land mines. Most interviews were conducted in French through a Khmer interpreter. We also interviewed Cambodian and foreign doctors and relief workers about the incidence of mine casualties[4] and the medical care available to both combatants and noncombatants. Cambodian government troops were also

[4]Casualty is defined as a trauma victim who either dies before reaching a clinic or hospital or requires admission for more than one day.

interviewed about their use of land mines and their knowledge of strategies used by the resistance forces.

In Thailand, we visited Site II, the largest of the Cambodian refugee camps. We interviewed refugees[5], including mine victims, about their experiences with mines in their home villages and during their flight to the border. Anonymous interviews were also conducted with soldiers in the three resistance armies about the strategic use of mines and their perceptions of the magnitude of mine dissemination in Cambodia. In these interviews, we showed resistance fighters photographs or illustrations of mines[6] and asked them to identify them as a way of determining the types of mines now being used in Cambodia.

This report documents the pervasive and indiscriminate use of land mines by the four warring factions in Cambodia. It examines the devastating medical, social, and psychological effects these weapons have had on Cambodia's civilian population and assesses their long-term effects on post-war rehabilitation. It also looks at the prospects for the organized repatriation of hundreds of thousands of Cambodian refugees from the Thai border camps, and the potential danger mines would pose to them on their return to Cambodia. Finally, the report recommends several measures to end the dissemination of mines in Cambodia, begin a mine survey, and implement an effective mine eradication program.

---

[5]Throughout this report the term "refugees" is used in its broadest sense, namely to refer to all Cambodians who have involuntarily crossed into Thailand since 1975. The Thai government refers to Cambodian refugees as either "illegal aliens" or "displaced persons."

[6]"Dummy" cards, illustrations of mines known not to be deployed in Cambodia, were included as a secondary check.

II. BACKGROUND AND HISTORY OF MINE WARFARE

Unlike bombs or artillery shells, which are designed to explode when they approach or hit their target, land mines lie dormant until a person, a vehicle, or an animal triggers their firing mechanism. They are blind weapons that cannot distinguish between the footfall of a soldier or that of an old woman gathering firewood. They recognize no cease-fire, and long after the fighting has stopped, they can maim or kill the children and grandchildren of the soldiers who laid them.

Land mines were developed soon after World War I as a response to the tank.[7] Just as the refinement of the internal combustion engine fostered the development of the tank as a counter to the stalemate of trench warfare, the invention, in the 1920s, of the easy-to-handle, powerful, and lightweight explosive trinitrotoluene (TNT) led to the development of the first reliable anti-tank pressure mines. During World War II, these flat steel cylinders, measuring about 30 cm in diameter and containing about 10 kg of TNT, were used extensively by all sides. Anti-tank mines had one major weakness: they could be easily removed by the enemy, who would plant them in their own minefields.

To keep mine-clearing soldiers at bay, both German and Allied troops began "seeding" their anti-tank minefields with small metallic or glass containers containing a pound or less of explosive. These early anti-personnel mines were activated by the direct pressure of 15 to 40 pounds on pins projecting from the mine, or by a few pounds of pull on a trip wire. One of the most

---

[7]See J.C. Beyer, *Wound Ballistics* (Washington, D.C.: Office of the Surgeon General, Department of the Army, 1962), p. 32.

effective of these mines, a German-made "bounding" device, popped up from the ground to a height of seven feet before exploding, shooting hundreds of steel balls half the length of a football field.

Soldiers also booby-trapped anti-tank mines to prevent their removal. In the early stages of the war, most of these devices were improvised with hand grenades or simple electric fuses. Later, more complex machine-made fuses were rigged to an explosive charge that would easily detonate when pressure was applied or when an electrical circuit was closed.

It was not long before improvised explosive devices (IEDs) and anti-personnel mines were being used as weapons in their own right, rather than merely to protect anti-tank mines. Both weapons were used to demoralize troops or terrorize civilians. Japanese soldiers, for instance, often booby-trapped harmless, everyday objects, such as pipes, flashlights, radios, and fruit cans. The practice of booby-trapping the bodies of dead or wounded soldiers, although officially denied, was also common.

Advances in mine warfare, as in all areas of weaponry, accelerated in the decades following World War II, primarily in response to changing battlefield requirements and the development of new military technologies. In the early 1960s, the United States first introduced the use of a new and sophisticated class of contact anti-personnel mines, known as remotely delivered mines or "scatterables,"[8] to stop the flow of men and material

---

[8]Scatterables are land mines that can be deployed from aircraft or launched by mortar or artillery and should not be confused with the more recent innovation, the "cluster bomb." Cluster bombs are deployed from aircraft and are designed to explode before reaching the ground. However, because of faulty fuse design or incorrect deployment, large numbers of these bombs do not explode and become, in effect, anti-personnel mines by default. Scatterables are land mines by design that

from North to South Vietnam through Laos and Cambodia. The most commonly deployed were the BLU 43 and 44, nicknamed "dragon tooth" because of their needlelike shape. American pilots dropped so many of these mines they referred to them as "garbage" because they could be scattered from the air and landed on the ground without detonating. When stepped on, the device, which weighed only 20 grams, could tear off a foot. (The BLU 43 and 44 were the forerunners of the Soviet PFM-1 "Butterfly" mine used extensively in Afghanistan.) Another remotely-delivered mine widely deployed by the United States in Vietnam was the BLU 42, or "spider" mine, which sent out eight trip wires, like spider legs, after landing on the ground.

For all their tactical advantages, scatterable mines had their drawbacks. Because of the hit-and-run nature of the Vietnam war, American ground forces often found themselves retreating through areas that their own pilots had saturated with mines days, or even hours, before. Another problem was the limitation on the accuracy with which the mines could be delivered and confined to a designated area. Much depended on the availability of sophisticated navigational and weapon-guidance systems, user competence, terrain, and the prevailing conditions of combat and weather. The indiscriminate nature of remotely delivered mines meant that they endangered civilians who passed through a mine-saturated area during and after skirmishes. Since the increase in civilian conflicts beginning in the 1970s, the land mine, like the automatic rifle, has become a weapon of choice for many armies and resistance groups around the world. Not only are they durable and effective, but they are readily available from both governments and the vast global network of private arms suppliers. Mines are also easy and relatively cheap to manufacture locally. The 1989 edition of *Jane's Military Vehicles and Logistics*[9]

---

can be deployed from aircraft or launched by mortar or artillery.

[9]Published by Jane's Information Group, Coulsdon, Surrey, UK.

lists 76 pages of different types of land mines in use by the major armed forces of the world, and the list is by no means comprehensive. As scientists invent new high-technology devices, the old, equally lethal models are unloaded onto the surplus arms market[10] or supplied directly to armies or guerrilla groups, usually in developing countries.

Advances in land mine design do not mean that anti-personnel mines are now more effective than earlier models. Because the mines of the new generation of mines contain powerful composite explosives and have extremely low metallic signatures, they are often extremely difficult to detect and deactivate or destroy. As a result, they can remain a threat to civilians long after the cessation of hostilities.

Mines commonly kill or inflict ravaging wounds, usually resulting in traumatic or surgical amputation, because the victim is usually very close to the detonation, often standing directly over the mine. If other people are in the area, they, too, can be killed or injured. Mines are often so sensitive they can be detonated by nearby detonations. Thus, the activation of a single mine may fire more mines in the vicinity.

Medical studies of combatants injured by land mines and other munitions indicate that early evacuation from the battlefield and prompt surgical care is crucial to saving lives and reducing disabilities. In Vietnam (1965-1973) and Lebanon (1982), medical facilities run by the U.S. and Israeli military respectively achieved

---

[10]For instance, in 1989, mine eradication teams in Paktia Province in Afghanistan found British MK.7 anti-tank mines with the sale lot numbers painted in red on their bases.

treatment results previously unsurpassed in war surgery.[11] This was due to the short transportation distances, the availability of helicopters, and well-equipped medical facilities. In most conflicts, however, battlefield first aid, evacuation, and treatment facilities are far from ideal, with resultant high morbidity and mortality.

Military personnel injured by land mines stand a better chance of receiving prompt medical care than civilians. To begin with, foot soldiers usually travel in groups and carry first-aid equipment. They can also radio to military bases or camps for transport and further medical assistance. In contrast, few, if any, civilians caught in or near war zones have access to rapid evacuation facilities.

Those civilians most likely to encounter mines are the rural poor who live far from towns or cities with proper medical facilities. Peasants foraging for wood and food or tilling their fields are particularly at risk.[12] Children herding livestock are likewise extremely vulnerable--the children regularly take their charges to fresh pastures and thus traverse wide tracks of untrodden land. The solitary nature of shepherding means the wounded child will often die slowly and painfully from the combined effects of blood loss and exposure. Sometimes children, attracted by the unfamiliar and unaware of the hazard, play with mines, with devastating results.

Even when civilians injured by mines reach medical facilities, they often fail to receive proper care because supplies of

---

[11]See L.D. Danon, et al., "Primary Treatment of Battle Casualties in the Lebanon War," *Israeli Journal of Medical Science*, 20 (1982):300-302, and R.M. Hardaway, "Viet Nam Wound Analysis," *Journal of Trauma*, 18 (1978):635-643.

[12]See, Mines Advisory Group, *Report of the Afghanistan Mines Survey*, March 1991.

X-ray film, anesthetics, surgical equipment, and antibiotics are unavailable or in short supply. Land-mine victims are also more likely to require amputations than victims wounded by other munitions. For instance, a study of Afghan war casualties admitted to a Pakistani border hospital from 1985 to 1987 revealed that 73 percent of land-mine injuries resulted in amputation. Of injuries from other fragmentation weapons, 18 percent required amputation, while only two percent of firearm injuries resulted in amputation.[13] In many cases, amputation is required because those aiding the mine victim fail to loosen tourniquets on the wounded limbs at regular intervals.

Land Mines in International Law

Land mines[14], unlike chemical and biological weapons, have never been banned. On the contrary, international law specifically permits the use of land mines to achieve military objectives. However, the 1981 Protocol on Prohibitions or Restrictions on the Use of Mines, Booby Traps, and Other Devices, otherwise known as the Land Mines Protocol, does contain restrictions on mine warfare which are designed to protect

---

[13]See J. Rautio and P. Paavolainen, "Afghan War Wounded: Experience with 200 Cases," *Journal of Trauma* 28;4 (1988):523-525. Also see, for example, D.E. Johnson et al, "Epidemiology of Combat Casualties in Thailand," *Journal of Trauma* 21:6 (1981):486-488 and D.B. Adams and C. W. Schwab, "Twenty-one-year Experience with land Mine Injuries," *Journal of Trauma* 28;1(Supplement) (1988):S159-162.

[14]Mines are as defined as "any munition placed under, on or near the ground or other surface area and designed to be detonated or exploded by the presence, proximity or contact of a person or vehicle. . . ." Land Mines Protocol, Article 2(1).

civilians.[15] The Land Mines Protocol is not directly applicable to internal armed conflicts, but many of its provisions are already a part of customary international humanitarian law and thus binding on the parties to the Cambodian conflict. The two key provisions in this regard are prohibition on indiscriminate use of mines and the obligation to minimize or avoid civilian casualties.

The Basic Rule: Protecting Civilians and Civilian Objects

Under customary law, civilians and civilian objects may not be attacked. U.N. General Assembly Resolution 2444, *Respect for Human Rights in Armed Conflict*,[16] adopted by unanimous vote on December 18, 1969, recognizes several principles of customary law protecting civilians. It states in part:

---

[15] This protocol, known as Protocol II, is one of three protocols annexed to the 1981 United Nations Convention on Prohibitions or Restrictions on the Use of Certain Conventional Weapons Which May Be Deemed To Be Excessively Injurious or To Have Indiscriminate Effects, UN Doc.A/Conf.95/15 (1980) ("UN Convention") (see Appendix A). It applies only to international armed conflicts and to some self-determination wars. By 1989, only 32 countries had ratified, accepted, approved or acceded to the UN Convention, including two of the major mines suppliers in the Cambodian conflict, the Soviet Union and China. Both the USSR and China consented to be bound by the protocols. The Soviet Union and Vietnam, which signed but did not ratify the UN Convention, supply mines to the Phnom Penh government. China is the main supplier of mines to the resistance forces, particularly the Khmer Rouge. For its part, the United States, which has signed but not ratified the UN Convention, has supplied mines to the non-communist guerrilla armies of Prince Sihanouk and the Khmer People's National Liberation Front (KPNLF). Cambodia has not signed or ratified the UN Convention.

[16] See second and third paragraphs in the preamble of G.A.Res. 2444, 23 U.N. GAOR Supp. (Wo. 18) p. 164, U.N. Doc. A/7433 (1968).

a) that the right of the parties to a conflict to adopt means of injuring the enemy is not unlimited;

b) that it is prohibited to launch attacks against the civilian population as such;

c) that a distinction must be made at all times between persons taking part in the hostilities and members of the civilian population to the effect that the latter be spared as much as possible . . . .

The Land Mines Protocol was adopted largely in response to the large number of civilian casualties caused by mines and unexploded munitions in Vietnam. It derives its provisions from customary law principles, and among other things, requires that combatants take "feasible precautions" (defined as "practicable or practically possible") under the circumstances to protect civilians from the effects of mines and booby traps.[17] The parties are required to keep records of minefields so that they can be cleared once hostilities have ended.[18] It prohibits in all circumstances the use of mines "either in offence, defence or by way of reprisals, against the civilian population as such or against individual civilians." [19]It also prohibits the use of land mines "in any city, town, village or other area" where civilians are concentrated, unless combat between ground forces is taking place or imminent in the area and the mines are placed 1) around a military

---

17 Land Mines Protocol, Article 3(4).

18 Land Mines Protocol, Article 7(1).

19 Land Mines Protocol, Article 3(2).

objective,[20] or measures such as putting up warnings are taken to protect civilians from the effects.[21] Scatterable or remotely delivered mines, defined as any mine "delivered by artillery, rocket, mortar or similar means or dropped from an aircraft," are specifically banned unless their location can be accurately recorded or they contain mechanisms to render them harmless "when it is anticipated" that the mines no longer serve a military purpose.[22] The Protocol also requires that "effective advance warning" be given to civilians before scatterable mines are delivered or dropped "unless circumstances do not permit."[23]

Article 3(3) of the Protocol prohibits the indiscriminate use of land mines. It defines indiscriminate use as any placement of mines:

> (a) which is not on, or directed at, a military objective; or
>
> (b) which employs a method or means of delivery which cannot be directed at a specific military objective; or
>
> (c) which may be expected to cause incidental loss of civilian life, injury to civilians, damage to civilian

---

[20] A military objective is defined as "any object which by its nature, location, purpose or use makes an effective contribution to military action and whose total or partial destruction, capture or neutralization in the circumstances ruling at the time, offers a definite military advantage."Article 2(4), Protocol on Prohibitions or Restrictions on the Use of Mines, Booby Traps and Other Devices (Protocol II).

[21] Land Mines Protocol, Article 4(2).

[22] Land Mines Protocol, Article 5(1).

[23] Land Mines Protocol, Article 5(2).

objects, or a combination thereof, which would be excessive in relation to the concrete and direct military advantage anticipated.

If a weapon cannot with any reasonable assurance be directed at a military objective, it is considered "blind" and, under Article 3(3), indiscriminate. Contact land mines are blind when left in an area through which civilians pass, since they can be detonated by civilians as well as fighters. Experts on the laws of war also state that "land mines, laid without customary precautions, and which are unrecorded, unmarked, or which are not designated to destroy themselves within a reasonable time, may also be blind weapons in relation to time."[24]

Therefore a mine not programmed to self-destruct and not removed from an area after fighting there has ceased becomes blind or indiscriminate as well.

Prohibition of Disproportionate Attacks

The legitimacy of a military target does not provide unlimited license to attack it. The customary law principles of military necessity and humanity require that the attacking party always seek to avoid or minimize civilian casualties.

The Land Mines Protocol codifies the "rule of proportionality" in customary law as it relates to collateral civilian casualties and damage to civilian objects. It thus prohibits as indiscriminate any placement of mines "which may be expected to cause incidental loss of civilian life, injury to civilians, damage to civilian objects, or a combination thereof, which would be

---

[24]M. Bothe, K. Partsch, and W. Solf, *New Rules for Victims of Armed Conflicts: Commentary on the Two 1977 Protocols Additional to the Geneva Conventions of 1949 (Geneva: 1982) ("New Rules")*, p. 305.

excessive in relation to the concrete and direct military advantage anticipated."

This two-pronged test of proportionality requires an assessment of the "concrete and definite military advantage" expected: such an advantage should be "substantial and relatively close," and "advantages which are hardly perceptible and those which would only appear in the long term should be disregarded."[25] Under this test, the possibility that enemy troops may at some undefined time in the future move across a certain path may be too remote and unsubstantial to qualify as a "concrete and definite military advantage."

The second prong of this test is that the foreseeable injury to civilians not be "excessive" in relation to the expected military advantage. Excessive damage is a relational concept which requires a good-faith balancing of disparate probabilities, but there is never a justification for excessive civilian casualties.[26]

These two factors must be weighed in good faith by the commanders responsible for mining. In our opinion, the possible military advantage of injuring a guerrilla or deterring guerrilla travel by a mine laid on a footpath, without warnings and left for an indefinite time, is frequently not sufficiently "concrete and direct" to outweigh the likely injury to civilians. The objective of injuring a combatant will not be achieved if a civilian steps on the mine first.

---

[25]International Committee of the Red Cross, *Commentary on the Additional Protocols of 8 June 1977 to the Geneva Conventions of 12 August 1949 (Geneva: 1987) ("ICRC Commentary")*, p. 684.

[26]*ICRC Commentary*, pp. 625-26.

Under this customary law rule as well, the practice of leaving unmarked, unrecorded land mines that do not self-destruct in civilian-traveled areas is indiscriminate.

Prohibition Against Starvation of the Civilian Population

By prohibiting starvation of the civilian population as a method of warfare or combat, Article 54 of Protocol I and Article 14 of Protocol II of 1977 to the Geneva Conventions of 1949 establish a substantially new rule which has been accepted by many governments as customary law,[27] and which imposes important restrictions on the use of land mines, especially in farming areas.

Article 14, Protocol II provides:

> Starvation of civilians as a method of combat is prohibited. It is prohibited to attack, destroy, remove or render useless, for that purpose, objects indispensable to the survival of the civilian population, such as foodstuffs, agricultural areas for the production of foodstuffs, crops, livestock, drinking water installations and supplies and irrigation works.

This prohibits starvation as a method of combat, "i.e., when it is used as a weapon to destroy the civilian population."[28]

While recognizing that it is still permissible to starve the enemy army, the article imposes sharp limits on that practice. That

---

[27] **See** Charles A. Allen, "Civilian Starvation and Relief During Armed Conflict: The Modern Humanitarian Law," 19 ***Georgia Journal of International Law and Comparative Law*** 1 (1989).

[28] *ICRC Commentary*, p. 1458.

"objects indispensible for the survival of the civilian population" may also be of benefit to the enemy army does not give license to attack them. The narrow exception is where the objects are specifically intended as provisions for combatants,[29] which is generally taken to mean foodstuffs actually in the hands of the enemy armed forces. It is not permitted to destroy or render useless agricultural areas for the production of foodstuffs because as a practical matter it is impossible to distinguish between the part intended for military and that intended for civilian use.[30]

Crops and agricultural fields may be attacked, however, when used in direct support of military action; the *ICRC Commentary* provides an example:

> What is the position if such objects hinder the enemy in observation or attack? This might be the case if crops were very tall and were suitable for concealment in a combat zone. . . . [I]f the objects are used for military purposes by the adversary, they may become a military objective and it cannot be ruled out that they may have to be destroyed in exceptional cases, although always provided that such action does not risk reducing the civilian population to a state of starvation.[31]

Neither party may destroy objects indispensable to the survival of civilians because it suspects those civilians of supporting the adversary. This is the rule regardless of whether the civilians

---

[29] *Id.*, p. 1458.

[30] *New Rules*, p. 340 (commenting on Article 54, Protocol I).

[31] *ICRC Commentary*, p. 1459 (footnote omitted).

live in territory controlled by that party or its adversary.[32] The *ICRC Commentary* notes:

> To deprive the civilian population of objects indispensable to its survival usually results in such a population moving elsewhere as it has no other recourse than to flee. Such movements are provoked by the use of starvation, which is in such cases equivalent to the use of force.[33]

Thus the counterinsurgency tactic of "draining the sea" -- or forcing civilians to move away from the guerrillas who live off them -- by means of depriving the civilians of food or rendering their fields useless for cultivation is prohibited by this article.

The article points out the most usual ways in which starvation is brought about but the list is not exhaustive. The words "attack, destroy, remove or render useless" are used to cover "all eventualities," including chemicals used to pollute water or defoliants used to destroy a forest, according to the *ICRC Commentary*.[34]

No less than chemicals or defoliants, use of contact land mines in agricultural areas or on paths to these fields has the effect of rendering the areas useless for food production, because no one will be able to plant there. The customary law prohibition against starvation of civilians as a method of combat forbids use of land mines to accomplish those ends.

---

32 *ICRC Commentary*, p. 1459.

33 *ICRC Commentary*, p. 1459.

34 *ICRC Commentary*, p. 1458.

Recording Requirement

The Land Mines Protocol contains a recording requirement in article 7(1)(a). The provision states that "[t]he parties to a conflict shall record the location of . . . all pre-planned mine fields laid by them."[35] Although the Land Mines Protocol does not define the term "pre-planned," an authority notes:

> Since 'pre-planned' means more than 'planned,' a 'pre-planned' minefield is, by its nature, one for which a detailed military plan exists considerably in advance of the proposed date of execution. Naturally, such a detailed military plan could not exist for the vast majority of minefields placed during wartime. In the heat of combat many minefields will be created to meet immediate battlefield contingencies with little 'planning' or 'pre-planning.'[36]

The provision for recording is designed to facilitate removal at the end of the conflict, primarily for the benefit of civilians. Thus, at the cessation of active hostilities, the parties are to "take all necessary and appropriate measures, including the use of such records, to protect civilians from the effects of minefields, mines, and booby traps."[37]

---

[35] Land Mines Protocol, Article 7(1)(a).

[36] Burrus Carnahan, "The Law of Mine Warfare: Protocol II to the United Nations Convention on Certain Conventional Weapons," 105 *Military Law Review* 73,82 (1984). The recording requirement applies only to the location of pre-planned minefields, not to the location of individual mines therein, or to the composition or configuration of the mines within the field.

[37] Land Mines Protocol, Article 7 (3) (a).

Use of Plastic in Mines

Another protocol accompanying the 1981 UN Convention is the Protocol on Non-Detectable Fragments (Protocol I). It prohibits the use "of any weapon the primary effect of which is to injure by fragments which in the human body escape detection by X-rays." The use of plastic casing in mines, a common practice designed to make the mines non-detectable by metal detectors, is prohibited, because plastic shrapnel lodged in the human body when the mine detonates can not be located in radiographs.

III. TWO DECADES OF MINES

Mines have clearly been the weapon of choice of all parties in Cambodia since the Vietnamese invasion in 1979, but they were also used in the war at least a decade earlier. Cambodians are thus facing the task not just of clearing mines laid last year or the year before, but of finding and destroying mines laid regularly by different groups over the last 20 years. Many of those mines are now overgrown by vegetation, immersed in water or simply forgotten, but they may be no less lethal as a result.

The Vietnam War

The use of land mines in Cambodia dates back to the Vietnam War. In 1967, at the height of the war, the North Vietnamese opened a front on the Cambodian border and with the permission of then ruler Prince Norodom Sihanouk, established base camps in eastern Cambodia. To protect these installations, the Vietnamese placed land mines around their perimeters. In 1970, after the pro-American Prime Minister of Cambodia, Lon Nol, deposed Sihanouk in a coup, mine warfare moved into the interior of Cambodia as the Khmer Rouge and Vietnamese-backed Khmer resistance forces battled both the Lon Nol forces and each other. According to Cambodian doctor and writer Haing Ngor, Lon Nol troops "advancing toward huge, noisy enthusiastic Khmer Rouge meetings in the countryside would find only loudspeakers and a tape player -- and land mines buried around the tree that the sound equipment hung from."[38]

South Vietnamese forces joined Lon Nol's troops in attacking Vietcong bases in 1970; the North Vietnamese and

---

[38]Haing Ngor, *A Cambodian Odyssey* (New York: Warner Books, 1987), p.70.

Vietcong retaliated with attacks that reached beyond the eastern border to areas near Phnom Penh.[39] On April 30, 1970, without informing Lon Nol, the United States joined in with a massive bombing campaign aimed at destroying Vietcong sanctuaries in Cambodia.[40] Not only did the bombing fail to achieve that goal -- it also marked the beginning of a boost in support for the Khmer Rouge in the countryside. More Vietnamese troops poured into Cambodia and by August 1970, North Vietnam controlled most of western Cambodia from Battambang southward to the sea and the province of Kompong Speu south to Kampot.

From 1970 to mid-1972, the two major warring parties in Cambodia were the North Vietnamese and Lon Nol's army which eventually went down to defeat. The Paris Peace Accords of 1973, which ended U.S. involvement in the Vietnam War, helped seal the fate of Lon Nol's government and dissolved the tactical alliance between the North Vietnamese and Khmer Rouge. (Relations were already poor because of a series of raids and massacres of Vietnamese nationals by Khmer Rouge cadre in the early 1970's.) Now on their own, Khmer Rouge leaders organized Cambodians in the countryside into self-contained agricultural cooperatives. These areas were sealed off from the outside world

---

[39]Elizabeth Becker, *When the War Was Over: Cambodia's Revolution and the Voices of Its People* (New York: Simon and Schuster, 1986), p. 133.

[40]Even today, over 20 years later, bomb craters, some the size of baseball diamonds, still mar the Cambodian countryside, and peasants continue to be killed or injured by unexploded munitions. A significant fraction of these high-explosive munitions failed to explode, or were designed not to explode, on their initial use. See E. S. Martin and M. Hiebert, "Explosive Remnants of the Second Indochina War in Viet Nam and Laos," in A. H. Westing, *Explosive Remnants of War: Mitigating the Environmental Effects* (London: Taylor & Francis, 1985), pp. 39-49.

with boobytraps and patrolled around the clock, creating "miles-wide tracts of no-man's-land."[41]

From 1973 to 1974, the Khmer Rouge advanced toward the capital of Phnom Penh, mining areas they controlled along the Mekong River as they went. They finally reached Phnom Penh on April 17, 1975.

The Khmer Rouge

The break between the Khmer Rouge and Vietnamese led to tension along the Vietnamese-Cambodian border which broke out in fierce fighting by 1977. More fighting meant more mines. But by then, so many people were dying from other causes that mines seemed a minor factor. In the less than four years that the Khmer Rouge ruled, more than one million people died as a direct or indirect result of the government's policies. Records at Tuol Sleng, the former Phnom Penh secondary school turned interrogation center, indicate that more than 15,000 persons brought to the prison were executed, usually after being tortured. Only seven survived--prisoners whose technical skills were useful in running the death camp.

In the decade from 1968 to the Vietnamese invasion of Cambodia in December 1978, there was no respite from war to clear mines or stop their being planted. In early 1979, when all of Cambodia was in chaos and the Khmer Rouge in retreat, Haing Ngor writes that he decided to try and reach Phnom Penh from the northwestern town of Phum Phnom. Along route 5 heading towards the capital were

> mines, in holes next to the road and near the bridges. The rain had washed the layer of dirt from the metal detonating buttons, which were about the

---

[41]E. Becker, p. 164.

> size of kneecaps, so most of the mines were visible. Even so, an ox stepped on a mine, killing several people we had known from Phnum Ra and wounding others.[42]

It was worse near the Thai border:

> We walked cautiously around a bend and came upon the site of the [mine] explosion. It was a blood-spattered scene, an arm hanging from a tree branch, part of a leg caught in bamboo. Ten or more dead lay by the side of the path, and many more were wounded. I made a tourniquet, removed some large pieces of shrapnel from wound, tied makeshift bandages and advised the relatives on preventing infection. With no medical supplies there was little more to do. It was a terrible way to die, or to be maimed, after living through the Khmer Rouge and coming so close to freedom.
>
> The mines appeared on either side of the path, sometimes in the middle. They had coin-size detonator buttons, white or rusted in color. From the detonator buttons, trip lines made of nearly invisible white nylon thread led to tying-off points such as trees or rocks nearby....Whether the Vietnamese or the Khmer Rouge planted the mines didn't matter much to us. All we knew was that we had to keep our eyes on the trial, searching for white threads.[43]

---

[42]H. Ngor, p. 352.

[43] H. Ngor, p. 378.

As Vietnamese forces gained control of Cambodia and installed the People's Republic of Kampuchea (made up largely of former Khmer Rouge who had defected), the Khmer Rouge, together with the civilians they controlled, fled to Thailand. They were soon followed by people who were simply fleeing the fighting. Then, as 1979 progressed, famine forced tens of thousands of Cambodians to seek relief at the Thai border.[44] Thailand, fearing that it would be inundated by refugees, closed (and mined) its borders in March 1979. As a result, the refugees grouped together in a string of camps along the Cambodian side of the border.

In June 1979, the Royal Thai Army forced some 43,000 to 45,000 Cambodian refugees who had crossed into Thailand back into Cambodia. The refugees were rounded up from encampments in and around the Thai border town of Aranyaprathet and then transported by bus some 300 kilometers to a mountainous region on the northeastern border near the temple of Preah Vihear. From there they were forced to walk down the Dangrek escarpment, a mountainous and thickly forested ridge, back into Cambodia, without any directions on how to traverse the extensive minefield that lay at the foot of the escarpment. For days the operation went on. When a group of people tried to return, Thai soldiers opened fire on them.[45] Thousands of people died at Preah Vihear, mainly from dehydration, diarrhoea, and mines that both the Vietnamese and retreating Pol Pot troops had placed along the border.

---

[44]Actual food production for the 1979-80 season was only 538,000 tons compared to a requirement of 1,692,000 tons. See E. Mysliwiek, "Punishing the Poor: The International Isolation of Kampuchea," a report for Oxfam, United Kingdom, 1988, p. 25.

[45]See William Shawcross, *Quality of Mercy: Cambodia, Holocaust, and Modern Conscience* (New York: Simon and Schuster, 1984),pp 90-92.

Growth of the Resistance Forces

By late 1979, the international community, led by the United Nations Children's Fund (UNICEF) and the World Food Program (WFP), had set up a massive relief effort at the border, which, in turn, attracted more refugees. At the same time, various political and military groups had begun asserting control over the refugees. On the northern border, Ta Mok, a brutal Khmer Rouge commander responsible for the internal purges of 1976-78, and his army had taken control of several camps. Just south of Ta Mok's encampments, followers of Sihanouk had settled and formed the United Front for an Independent, Neutral, Peaceful and Cooperative Cambodia (FUNCINPEC). In the central border area, black marketeers and remnants of Lon Nol's government had established a lucrative trade in consumer goods from their encampments. Many of them would join forces in 1981 to form the non-Communist Khmer People's National Liberation Front (KPNLF) headed by former prime minister Son Sann. Further south, another Khmer Rouge commander, Son Sen, controlled a cluster of camps. And in the southern border area, adjacent to Thailand's Trat province, Pol Pot and several of his commanders were in charge.

All three factions maintained their own guerrilla armies. However, neither the Khmer Rouge nor the weaker KPNLF forces and the Sihanouk National Army (ANS), the military wing of FUNCINPEC commanded by Sihanouk's son, Prince Ranariddh, were a match for the Vietnamese, whose occupation army numbered 150,000 soldiers.

Democratic Kampuchea, the government of the Khmer Rouge, retained Cambodia's seat at the United Nations until 1982. During that time, the humanitarian aid provided to the border population permitted the Khmer Rouge to survive and rebuild.

Military aid was also provided, primarily by China, and from 1980 to 1986 by the United States.[46]

On June 22, 1982, in a meeting in Kuala Lumpur, Malaysia, Sihanouk signed an agreement with Son San and the Khmer Rouge leader Khieu Samphan to form the Coalition Government of Democratic Kampuchea (CGDK). Shortly thereafter, this tripartite coalition cobbled together -- with U.S. support -- by China and the Association of Southeast Asian Nations (ASEAN), took over Cambodia's seat in the United Nations. The CGDK was a tactical, if uneasy, alliance, a diplomatic fiction useful for opposing the Vietnamese installed People's Republic of Kampuchea (PRK). As a coalition, it neither exercised governmental authority inside Cambodia, nor did it have a headquarters or a common constitution. Sihanouk said of his partnership with his old foes, the Khmer Rouge: "We have to choose between letting the Vietnamese colonize Cambodia or working with the Khmer Rouge."

For Thailand, the CGDK and its resistance armies provided a convenient buffer against the Vietnamese. For the Chinese, support of the Khmer Rouge and, to a lesser extent, Sihanouk, was simply a continuation of its long-standing hostility towards its military enemy Vietnam and, by extension, the Soviet Union. For its part, the United States viewed its support of the CGDK as a means of isolating Vietnam and preventing recognition of the Phnom Penh government.

---

[46] From October 1986 to June 1991, the U.S. provided $20.3 million in non-lethal military assistance to the non-communist resistance groups under the International Security and Development Cooperation Act of 1985. See "Cambodia: AID's Management of Humanitarian Assistance Programs," United States General Accounting Office Report to Congressional Requesters, Washington, D.C., September 1991.

Until 1984, the war in Cambodia followed a seasonal pattern. During the rainy season, when heavy military equipment was, for the most part, useless, guerrillas from the border camps were able to infiltrate throughout Cambodia, spreading their propaganda, recruiting new fighters, and engaging in sabotage. During the dry season, Vietnamese troops, with help from the fledgling army of the People's Republic of Kampuchea, would push the guerrillas back to the border.

FIGURE 3.5

LOCATION OF UNBRO ASSISTED BORDER CAMPS
1985 - PRESENT

Current

- ● KHMER ROUGE
- ■ KPNLF
- ▲ FUNCINPEC

Former

○ ( ) Date of closure

Map courtesy of John Rogge and the Intertect Institute

In late 1984, as the northeast winds ushered in the dry season, the Vietnamese launched a major offensive against the guerrillas and their encampments on the Cambodian side of the border. For five months, they pounded the border with artillery fire, sending some 220,000 civilians and combatants fleeing into Thailand. The refugees gathered in 14 camps along the 700-kilometer border from Ubon in the north to Trat in the south. The Thai authorities agreed to grant them temporary asylum as "displaced persons," but only until security conditions permitted their return to Cambodia.

Nearly all the camps and their military appendages were located close to the border, partly to reinforce their temporary nature and partly to maintain their "buffer" function. With the guerrillas pushed into Thai territory, the Vietnamese effectively sealed off the border by laying extensive minefields. Similarly, the Thais and resistance forces, fearing a Vietnamese invasion, heavily mined key positions just over the border.

So dangerous was the trip across the minefields, few civilian refugees--no more than a few hundred, according to one source[47]--ventured back to Cambodia between 1985 and 1988. The mines also temporarily stopped the trans-border black market trade. However, for those willing to take the risk, there were "guides" who charged up to Thai Baht 200 (US$ 8) per person--an expensive proposition for most camp residents--to escort refugees through the minefields.

By late 1985, a year after the Vietnamese dry-season offensive, the refugee population inside Thailand had stabilized at about 300,000 persons, settling into several camps along the border (see Appendix B). Mines were reportedly placed around the camps to prevent Vietnamese incursions -- or to control the movement of refugees. Under the control of different resistance

---

[47]See John R. Rogge, "Return to Cambodia," p. 97-98.

factions, the refugees were periodically forced back across to border to carry ammunition or supplies needed by the guerrilla armies.

After the withdrawal of Vietnamese forces in September 1989, the movement across the border intensified, as each faction sought to establish a presence in the areas they had occupied. The Vietnamese withdrawal also gave the Khmer Rouge, by far the strongest of the three factions, a new lease on life, and led to intensified fighting between Phnom Penh government forces and Khmer Rouge guerrillas. Fear that the Khmer Rouge might return to power, combined with the changing international environment of the "post Cold War era", injected a dynamism into the peace process that produced the Perm 5 framework of September 1990.

Toward Peace

The first glimmer of hope that the end to the fighting might be in sight came on April 26, 1991, when representatives of the Phnom Penh government and the three resistance groups agreed to a cease-fire as of May 1. Two months later, in the Thai town of Pattaya, they agreed further to a moratorium on arms imports and to allow the 12-member Supreme National Council (SNC)[48], a transitional body formed in 1990 under a United Nation peace plan, to set up headquarters in Phnom Penh by the end of 1991.

In July 1991, at a meeting in Beijing, the factions further agreed that Cambodia's seat in the United Nations will be occupied by an SNC delegation led by Prince Norodom Sihanouk,

---

[48]Six of the 12 members of the Supreme National Council are representatives of the Phnom Penh government; each of the resistance groups has two representatives.

the exiled Cambodian leader.[49] They also announced an agreement to call for a special United Nations team to visit Cambodia and back up the cease-fire already in place and evaluate how to enforce the ban on outside military aid. Finally, in August 1991, in another round of meetings in Pattaya, the Phnom Penh government and three resistance factions agreed to reduce their military forces by 70 percent.

Under the provisions of the UN plan, until elections are held, the United Nations Transition Authority in Cambodia (UNTAC), will oversee key existing government ministries to ensure their political neutrality, and the Supreme National Council (SNC) will be both the institutional embodiment of Cambodian sovereignty and an advisory body to UNTAC, with some decision-making powers. The plan fixes no date for elections and makes only the abstract commitment to hold them after a long-term U.N. rule.

Nineteenth in a line of United Nations peace-keeping forces and missions dating back to 1948, UNTAC would have an unprecedented administrative role in Cambodia. It would also be the largest and most expensive peace-keeping force in the history of the United Nations. The agency would be comprised of both military and civilian personnel, numbering in the tens of thousands. With the SNC as an adviser, UNTAC would supervise Cambodia's ministries of foreign affairs, interior, finance, defense, and information. Its military forces would be responsible for monitoring the cease-fire and the cessation of the flow of foreign arms supplies to all factions, and supervising the partial cantonment and disarmament of all armed forces.

---

[49]The delegation would also include two delegates, Hun Sen and Hor Nam Hong, from the Phnom Penh government, as well as Khieu Samphan, a leader of the Khmer Rouge. See S. WuDunn, "Cambodian Factions Agree to Share Seat at U.N.," *New York Times*, July 18, 1991, p. A7.

That plan may be implemented some time in the future. But the menacing presence of hundreds of thousands of land mines is a very real problem for Cambodia now. In May 1991, a U.N. fact-finding mission that travelled to "liberated zones" held by the KPNLF and ANS in northwest Cambodia found:

> There was a heavy concentration of mines around military bases, water sources and abandoned villages. No mapping or recording of mines or mine incidents had been done. Villagers were aware of mines, but villagers continued to be killed or injured by them.[50]

Mines in Cambodia's interior and especially along its 700-kilometer border with Thailand are so widespread, they now rank, along with malaria and tuberculosis, as one of the country's three most serious public health hazards. Those at greatest risk from mines are returning refugees, the internally displaced, and all Cambodians living near the border.

---

[50]See "Report of UN Fact-Finding Mission to KPNLF and FUNCINPEC Areas of Cambodia (18-22 May 1991)," Office of the Special Representative of the Secretary-General for Co-ordination of Cambodian Humanitarian Assistance Programmes, June 3, 1991, p. 7.

IV. MINE WARFARE TODAY

For all parties to the Cambodian conflict, the main purpose of laying land mines has been to limit military operations by enemy forces: to deny the opposite side access to bridges, roads or strategic installations or to protect one's own forces from attack. The Phnom Penh government and the resistance groups have also used mines as offensive weapons. Government troops have placed mines around the perimeters of enemy villages and then bombarded them with artillery fire so that the "enemy" is forced to flee into minefields. The Khmer Rouge has used mines to channel and control the movement of people in the areas they control. Thai and Khmer black marketeers, often with the blessing of resistance forces, have used mines to secure their trade routes. Whatever the intended target, it is the sheer number of land mines and their indiscriminate use by all sides in the conflict which has resulted in tens of thousands of civilian deaths and injuries since the war began.

The Phnom Penh Government

The Phnom Penh government and, prior to their withdrawal, the Vietnamese forces, have used a two-pronged strategy: (1) the mass dissemination of mines, primarily to secure Cambodia's borders from incursions by resistance forces; and (2) the selective defense of key military and civilian installations and strong points. Because of where and how these mines are placed, even though for military purposes, they have posed an unacceptably high risk to the civilian population.

The Vietnamese laid mines in Cambodia in mid-1979, as they pushed the fleeing Khmer Rouge across the border into Thailand. It appears, however, that no attempt was ever made to record the location of these minefields, and Phnom Penh military sources claim that no maps have been handed over to the government by the Vietnamese Army. One key strategy used by the Vietnamese forces was to saturate the roads, trails and ground surrounding their bases and forward posts with mines. Anti-personnel mines were also used to channel and slow down resistance forces, often using inter-linked tripwire devices such as the POMZ-2, a Soviet-made anti-personnel mine. But the most damaging strategy introduced by the Vietnamese was the use of mines to isolate insurgents from their supporters in the villages by mining rice paddies and the margins of forests. The Phnom Penh forces have continued to use these strategies since the Vietnamese withdrawal in 1989.

Although there are no reliable records on the perimeters of minefields and their contents along Cambodia's border with Thailand, both the incidence and geographic spread of land mine casualties would suggest that there are hundreds of thousands of mines in the area. As of 1991, both government and resistance forces were still establishing new minefields, often overlaying existing ones. Even the term "minefield" has no real relevance in the border areas because it indicates a formality that does not exist.

The high density of mines along the border can be inferred from the reported numbers of mines removed in the Thmar Puok area by clearance teams funded by the U.S. Agency for International Development. They said in an interview with our delegation in April 1991 that teams had dealt with 6,000 anti-personnel devices in a one-kilometer stretch of ground close to an old Vietnamese military post and 3,800 mines in another two-kilometer section.

Government troops have used mines to protect key installations, especially bridges, from sabotage. There are hundreds of bridges over seven meters on the roads of Cambodia, all of which are prime targets for resistance forces. Foreign relief workers told us that government forces post skull-and-crossbones warnings and sometimes the type of mine on bridges. However, we saw no such warnings on bridges along the 315-kilometer highway that connects Mongol Borei in the northwest of the country to Phnom Penh, nor did we see warnings on bridges on the two 60-kilometer highways connecting the capital to Takeo in the south. Instead, government soldiers or militia were posted at most bridges, and the perimeters of mined areas were delineated by wooden poles and barbed wire entanglements which, when the bridge is over a river or watercourse, continues above and below the water surface.[51] The use of mines in this manner is, by nature and intent, responsible and does not constitute an unacceptable threat to non-combatants. The mines can be cleared with minimal risk to the surrounding civilian population.

According to soldiers we interviewed, troops lay mines along some roads at dusk and remove them at daylight. This tactic is often used on sections of road that government troops do not control at night. Although troops may not mine road surfaces, they heavily mine road verges and surrounding land up to ten meters on either side of the road. Anti-tank mines and directional devices such as the MON-100 are often used on verges and areas adjacent to the road. It is standard government practice to booby-trap anti-tank mines, usually by linking them to anti-personnel devices or grenades.

Government troops also use mines to limit contact between insurgents and villagers. This objective is usually achieved through the random deployment of anti-personnel devices, such as the

---

[51]Government forces employ similar defenses in the vicinity of railway bridges. Although hospital casualties indicate that some random use of mines on rail embankments may occur, there is no conclusive evidence that suggests which combatant faction is responsible for these incidents.

PMN-2 and POMZ-2, on the edges of forests close to towns and villages and in other areas where clandestine meetings may be likely to take place. Strong points under siege or threat of attack are also heavily mined, leaving only key access routes clear (these are routinely mined during hours of darkness and re-opened during daylight). These mine concentrations make agricultural and other land in the vicinity totally inaccessible.

In Svay Chek, for example, a village near Cambodia's border with Thailand, mines have virtually replaced crops in the surrounding fields and rice paddies. Svay Chek is situated in a key tactical position on the Sisophon-Thmar Puok road close to the Cheng (or Svay Chek) River. In December 1989, KPNLF forces took control of Svay Chek, which, at the time, had 15,000 inhabitants. During the siege and occupation of the village, KPNLF fighters were kept well supplied from their "liberated zones" less than 25 kilometers away.[52]

In February 1990, government troops shelled the village, after surrounding its perimeter with land mines. In the ensuing battle, hundreds of civilians were killed or wounded. By March, Svay Chek was back under government control. Since then, government troops have been able to keep their soldiers well-supplied from Sisophon and have considerable firepower within the village. But the government's greatest deterrence, according to KPNLF forces, is the wide belt of mines that surrounds the village.[53]

---

[52]U.S. support of the liberated zones and its funding of road construction operations has strengthened the KPNLF's logistic support chain to their forces at Svay Chek, as well as enabling them to rotate personnel on a regular basis.

[53]Government forces have deployed MON-100 and MON-50 directional mines along routes into Svay Chek. They have also laid several types of mines, including PMN, PMN-2, MBV 78/A2, POMZ-2,

By all accounts, neither the Phnom Penh government nor the resistance forces have kept systematic records of mined areas. A Red Cross worker in Battambang, Cambodia, put it this way:

> The problem with mines here in Cambodia, in comparison to other areas, is that they are the most important weapon. They are used as offensive weapons, as aggressive weapons....Here there is not a lot of direct confrontation. What you have is a concentration of soldiers in one place, the resistance will circle the area with mines then retreat and shell the area so the soldiers flee through the mined area....There are no maps because mines are given to soldiers like bullets.[54]

In a food stall in the bustling Battambang market, a group of government soldiers, dressed in uniform, spoke to our delegation about what they thought about the use of mines.[55] As one of the soldiers translated for the others, they responded to our questions as follows:

Q: Do you use mines?

A: Everybody uses mines. Most of us are farmers, and we know that we are causing a problem that makes the land

---

and grenades (the latter two are sometimes deployed with tripwires), around the village. KPNLF forces have placed anti-personnel mines--M14, Type 69 bounding mines, PMN-2, and some anti-tank mines--in front of their positions.

[54]Interview, Battambang, Cambodia, April 11, 1991.

[55]The soldiers also identified several mines (TM46, PMN, PMN-2, MD82B, M14, T-69, and POMZ-2) from photographs and drawings.

dangerous. In the future (laughs) the war will go, but the war will stay.

Q: Is that a Khmer saying?

A: It is our saying--a farmer's saying. When we fight mines are good weapons, unless, of course, I am blown up or my friends are blown up. But they are bad weapons, too. Bad for farmers and our families.

Q: Would you agree with a program to get rid of the mines? Would you see that as a good thing for Cambodia?

A: Yes, but who would be so crazy? Who would do a job like that?

Q: Would you help?

A: Maybe, but it will be more dangerous than the fighting. Anyway, who would pay us?

Q: Where do you lay most of the mines?

A: We put them around Pol Potist[56] (resistance) positions at night, and they do the same to us. We also use them in the forests and in places where we think the enemy will go. Also, when we find their (resistance forces) minefields, we sometimes put our mines among them, which gives them problems if they try to pick up their mines to use in other places.

Q: Do you mark the mines you place?

---

[56]Soldiers and civilian supporters of the Phnom Penh government often refer to all three guerrilla armies as "Pol Pot."

A: We use a death (skull and crossbones) sign at bridges and other important places. Sometimes we put one of the mines on show....other times we don't mark. Why should we? Then the enemy would know about the mines and would avoid them.

Resistance Forces

The Phnom Penh government, as well as the three resistance forces, have all been heavily influenced by foreign military training. Khmer Rouge mine strategy tends to mirror Chinese doctrine, while the strategy of the KPNLF and ANS--the two non-communist resistance forces--reflects British special forces training. Chinese training of the Khmer Rouge and the British/Thai Junior Commander Course attended by KPNLF and ANS officers until 1989 have devoted considerable sections of the curriculum to the use of improvised explosive devices and booby-traps. Classes range from improvised use of conventional munitions such as mines, through use of in-service and locally fabricated switches as initiators of explosive traps on to the use of commercial products to manufacture explosives.

These Junior Commander courses, each lasting six months, were conducted from 1986 to 1989 at a Thai military facility believed to be near the Burmese border. At least six courses are known to have taken place. Instruction was carried out by a uniformed British Army team drawn from the Special Air Service(SAS)[57] and the Royal Thai Army. Each course consisted

---

[57] The British Government consistently denied allegations that the British Forces were involved in training the Cambodian resistance. For instance, in October 1990, then Prime Minister, Margaret Thatcher said, "There is no British Government involvement of any kind in training, equipping or co-operating with the Khmer Rouge forces or those allied to them." Then, on June 27, in a written parliamentary answer, the Armed Forces Minister, Archie Hamilton, admitted that British Forces "provided training to the armed forces of the Cambodian non-communist

of 50 students, 25 each from the ANS and KPNLF, who were selected on the basis of their physical fitness and weapons knowledge.

In April 1991, our delegation interviewed several resistance fighters who had attended the courses. We also examined their course notes and other documents pertaining to the courses. According to these informants, the training was conducted in strict secrecy: students were not told where they were being taken to and were only allowed outside the camp during training exercises. Thai soldiers taught them unarmed combat and physical training, while seven British instructors[58] taught them about tactics, mine warfare, demolition, weapons, navigation, first aid, video, radio, and communications.

Our informants said that the overall objective of the course was to produce effective field commanders who could operate independently in enemy territory, with emphasis on the destruction of the military and civilian infrastructure. The students were taught practical command skills required in anti-insurgency operations, including the use of minefields as a planned and hasty defence. For three months during the course, the students split into two groups: one specializing in tactics, the other in demolition. Our informants had specialized in the latter and thus were unable to provide information about the tactics training.

Students in the demolition group were taught to be expert in the use of all explosives and, when no regular explosives were

---

resistance from 1983 to 1989."

[58]According to our informants, the British instructors wore uniforms and carried sidearms. The informants drew the "winged dagger," the SAS cloth badge, and correctly described the fawn-colored beret worn by SAS troopers.

available, how to manufacture them. They were taught all the major fuses, switches, and detonators used worldwide, as well as those used for industrial purposes. Detailed instruction was given in standard demolition procedures and peripheral subjects, such as the use of conventional ordnance as emergency demolition charges. Students were trained to calculate and design the charges necessary to destroy both military and civilian targets. These included aircraft, railways, trains, bridges, power stations, office buildings, armored vehicles, artillery, missiles, and other key targets.

Our informants said that a considerable amount of time was spent on the use of improvised explosive devices, booby traps, and the manufacture and use of time-delay fuses. They said a British instructor taught their class how to draw a map of a minefield but explained that "such maps are rarely drawn and it's hardly practical to bother with them."

The Chinese provide training to all the resistance factions but, as would be expected, given their political stance, the most comprehensive instruction is reserved for Khmer Rouge forces. Our delegation obtained a Chinese military manual used in the training of both KPNLF and Khmer Rouge fighters. The manual emphasizes the use of improvised booby-traps, often employing "over-kill" quantities of explosives for maximum physical and anti-morale effect. Sections in the manual relating to the demolition of infrastructures, particularly railroads and bridges, show that the training is of a technically sophisticated nature.

Several resistance fighters told us that in many parts of the border belt, their own mines were as great a threat to them as those left by the enemy. Ung Samuth, an amputee and former Khmer Rouge fighter, said no one in his unit ever kept records of where they had placed mines. "Now, all these years later, it will be difficult to locate the mines," he said, "because many of the men who laid them are dead."

In Site II, a KPNLF soldier described how he came to the realization that he and his fellow soldiers were laying mines in their own minefields:

> Nobody likes mines but everyone uses them. On my first active duty we had a leader who kept a notebook about where we put mines, all the experienced fighters laughed at that and asked "Who will read such a book?" I thought it was a good idea but only if others did the same thing and, of course they don't. Last year, we were ordered to put mines near a path. As we were doing this I had a feeling--you know--when you think you know a place, that you have been in the place before. Anyway, my attention was distracted because one of the others was blown up, then the man on my right shouted that we were in a minefield and started to walk back to the path and he stood on a mine. I was the only one who got out without being killed or injured. I moved slowly and carefully, using a stick to check the ground. I knew then why I knew the place, why I remembered it--I had put mines there only three months before. We were putting mines in our own minefield--it was crazy, killing our own fighters.

Resistance fighters also told us that they often bombard villages with mortars or artillery shells until the occupying troops or civilians retreat into the forest. The resistance fighters then advance into the village and lay mines before withdrawing. Several resistance fighters and noncombatants told us that water sources, access routes, and surrounding farmland are heavily mined (this is supported by evidence of casualties who have tried to return to villages). They also said that all sides booby-trapped dwellings and common buildings.

The Khmer Rouge seem to use mines in a somewhat more sophisticated manner, but no less indiscriminately, than the KPNLF and ANS. The Khmer Rouge often use mines to channel and control population movements. This suggests that they may have kept some records of minefields. However, there is also evidence that the Khmer Rouge randomly deploy mines in the same way as other factions in areas not under their control; this is particularly evident in Kampong Speu province.

The placement of larger, anti-tank mines along roads and railways is also attributed to the Khmer Rouge. Trains operating along the single functioning railway in Cambodia (from Kampot, via Phnom Penh, to Battambang) are preceded by two flat cars piled with timber in an attempt to trigger mines before damage is done to the locomotive. Most passengers ride in cars behind the locomotive. But for those who wish to risk their lives for a free ticket, there is always *gratis* seating on the front two cars.

During our visit in April 1991, we watched from Highway 5, which runs parallel to the railway line just north of Pursat, as a train passed. Indeed, there were passengers riding on the logs on the cars in front of the locomotive and, from the top of the car behind it, several government soldiers fired bursts from their automatic weapons to keep resistance fighters away.

All resistance factions convert grenades into mines by a simple tripwire device. They are used in jungle areas and close to river banks, normally hung from, or fastened to, tree branches several feet from the ground. They are initiated by tripwire at ground level.

One standard improvisation used by all factions is the multiple stacking of mines. This is done to hinder their removal by enemy forces or to increase the explosive effect and range of the mines. In its most extreme form, a tactic employed by Phnom Penh, an anti-tank mine called the TM-46 is placed above a standard pressure mine called the PMN with a Type 69 bounding mine at the bottom of the stack. The combination can be initiated in three ways. First, it can be detonated by a tank or heavy vehicle. Second, a person, exerting .23kg of pressure (the operating force required to detonate the PMN), can detonate the stack. At that pressure even a child can easily initiate the combination. This turns the TM-46 into an anti-personnel mine with an explosive content of more than five kilograms of TNT. Finally, the Type 69 mine can be detonated by a buried pull-wire attached to a hidden firing position. Thus, the combination can be remotely detonated, the PMN and TM-46 having first been projected upwards by the Type 69 device. A government soldier who had witnessed such an explosion told our delegation, "some of the [enemy] patrol simply disappeared."

Resistance forces, we found, routinely booby-trap mines to prevent their removal by placing a pressure-release device below the primary mine or linking it to another device, often a hand-grenade, by a hidden tripwire. The Khmer Rouge rely heavily on booby traps and their training programs place considerable emphasis on fairly complex traps, often adaptable to use with large quantities of explosive.

Land Mines and Other Devices in use in Cambodia

Given the extraordinary number of types of mines in use in Cambodia, it is impossible to give a comprehensive picture of all the mines used by both government and resistance forces. The following information, however, is thought to be indicative of the current situation and a realistic overview of mass mine-laying strategy, particularly in the border "belt," used by the Phnom Penh government and resistance forces.

The Phnom Penh Government

The types of mines being used by the Phnom Penh government are as follows:

PMN/PMN-2
Source: USSR
Manufacturer: Soviet State Arsenals
Type: Anti-personnel/blast
Initiation: Pressure

PMN

Both of these anti-personnel pressure-initiated blast devices, particularly the latter, have been used in great quantity throughout the Thai border belt. The devices are either buried or placed on the ground or other surfaces and covered with natural camouflage. Both mines have been used in conjunction with other devices in stacked configurations. The PMN has a built-in, 20-minute delay after arming, and can be easily by specialists. In the case of the PMN-2, the arming process is irreversible and there is no recognized neutralization technique. Both devices have a thermo-plastic outer casing.

POMZ-2/POMZ-2M
Source: USSR
Manufacturer: Soviet State Arsenals
Type: Anti-personnel /fragmentation
Initiation: Tripwire

This anti-personnel fragmentation mine has been used extensively and, in most cases, is linked to other explosive devices, such as ball-mines and grenades, to provide a wide killing area from a single contact. The POMZ-2M is a more modern version of the POMZ-2; the main difference is that it has five, rather than six, rows of fragmentation.

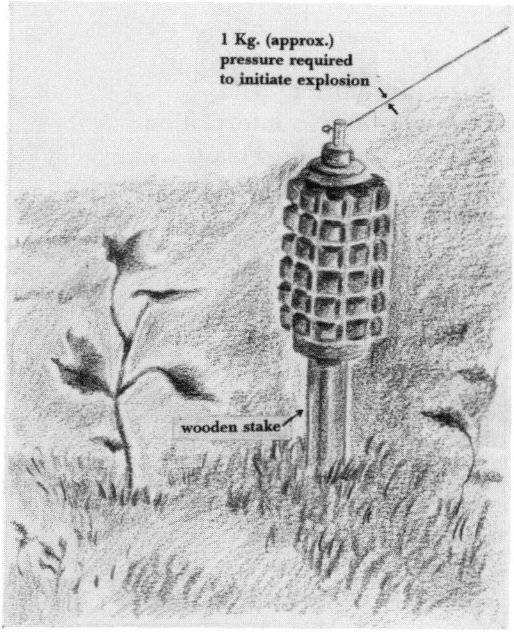

POMZ-2

Ball-Mine
Source: Vietnam
Manufacturer: Vietnam State Factories
Type: Anti-personnel/fragmentation
Initiation: Pressure or tripwire

There are two variations of this small ball-shaped pressure or tripwire-initiated device, often inter-linked with POMZ-2, widely used by government forces. The bright orange or red coloration of these devices makes them ideal as a calculated method of channelling enemy forces. The mine can be set to explode when stepped on or disturbed by tripwire.

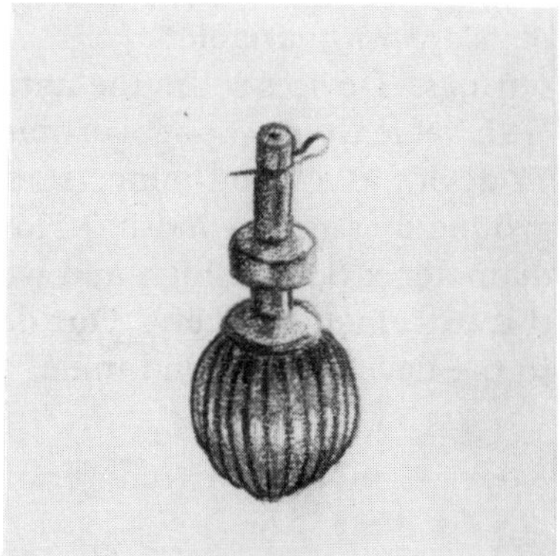
"BALL" MINE

Type 72
Source: Insufficient data
Manufacturer: No information
Type: Anti-personnel/blast
Initiation: Pressure or anti-disturbance

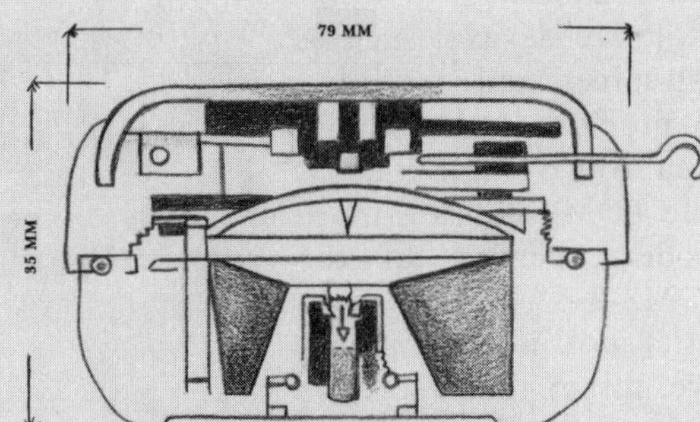

MODEL 72 ANTI-PERSONNEL MINE
SOURCE: CUTAWAY SKETCH FROM CHINESE TRAINING MANUAL ISSUED TO KHMER ROUGE TRAINEES

This small, plastic anti-personnel mine is used by both government and Khmer Rouge forces as both a pressure and anti-disturbance device. The latter mode is designed to ensure detonation when the device is handled or disturbed in any way, making it extremely unstable. We do not know whether there are two distinct devices or one device with variable settings. Devices set in the anti-disturbance role are often referred to as "72b." at least one nongovernmental organization has reported the Type-72 device as a "wind-mine," stating that the device can be blown over the ground by strong winds.[59] However, the device's size and weight (8 cm diameter x 3.5 cm high and weighing approximately 140 grams) make this extremely unlikely. Our delegation found no Khmers who referred to the device as a "wind-mine."

---

[59]See, for example, P. Carey, "Cambodia's Unending Agony," *The Independent Magazine*, August 4, 1990, p. 31.

"Duen" Mine
Source: Vietnam
Manufacturer: Insufficient data
Type: Anti-personnel
Initiation: Probably pressure or tripwire

Little is known about this Vietnamese device reportedly widely used by government troops. Of metal construction, the device is said to be "shaped like a bell," and is approximately 4.5 to 5 inches in diameter. Some sources say the device is locally improvised and varies in size.

TM-46
Source: USSR
Manufacturer: Soviet State Arsenals
Type: Anti-tank
Initiation: Pressure/tilt rod

This seems to be the most common anti-tank, or vehicle, mine in use by all factions in Cambodia. It is often buried up to one meter deep under a tree stump, thus making it difficult to detect while retaining effective pressure initiation when the vehicle passes over the trunk. Some reports indicate that the Phnom Penh government may also be deploying the TMN-46 mine. This device is almost identical to the TM-46, and can be booby-trapped with an MV-5 Fuse to prevent removal.

OZM-3
Source: USSR
Manufacturer: Soviet State Arsenals
Type: Anti-personnel bounding fragmentation (designated obstacle)
Initiation: Depends on fusing, as it can be remote, electrical, pressure, tension-release, or tripwire

This device is frequently referred to as a Type 72, causing some confusion with the anti-personnel and anti-tank mines of the same designation referred to elsewhere. This is an anti-personnel bounding device exploded by tripwire and said to be used in large numbers in some heavily forested areas of Cambodia. The mine has a lethal radius of twenty-five meters at a height of between 1.5 and 2.5 meters. It is also used by resistance factions, particularly the Khmer Rouge.

MON-50/MON-100
(Minnoye oskolochonym napravleniem)
Source: USSR
Manufacturer: Soviet State Arsenals
Type: Anti-personnel directional fragmentation
Initiation: Remote or tripwire

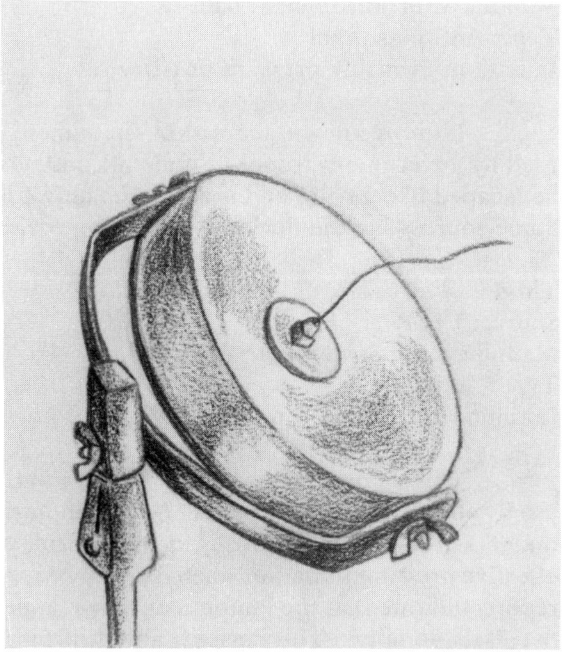

MON-100 DIRECTIONAL MINE

These two directional mines have been used on the border primarily as ambush and position-defence weapons and are normally fired remotely, singly, or in pairs. There are no reports indicating use of these devices in conjunction with tripwire, sonar, or other independent initiation devices. Standard government practice is to recover these mines in the case of withdrawal or when an ambush is not actuated. The MON-50 is a copy of the U.S.-manufactured M18A1 Claymore mine and projects fragments over 50 meters in a 60-degree arc. The MON-100 uses a 5-kg charge of TNT to project 450 steel fragments with an approximate lethal range of 100 meters.

Grenades

Grenades, usually MBV78/A2, are often used as mines, sometimes inter-linked with POMZ-2 and ball-mines. Government soldiers can often be seen carrying grenades with a length of tripwire attached to the pin and coiled ready for use. They may be deployed in this role as often as they are used in the conventional manner.

The Resistance Forces

The type of mines used by the resistance forces are as follows:

M16A1
Source: USA (Also Greece & India)
Manufacturer: USA
Greece - Hellenic Arms Industry SA
India - Ordnance Factory Board, Calcutta

This bounding anti-personnel mine is initiated by tripwire or pressure. Of American origin, it is used by both Khmer Rouge and KPNLF forces. The device, in the pressure mode, requires between 3.6 and 20 kilogram of downward force on one of three prongs protruding from the M605 combination fuse screwed into the top of the mine body. When set to explode by tripwire, between 1.6 and 3.8 kilogram of pressure is required. The mine reaches a height of approximately one meter before exploding and scattering metal fragments in all directions.

Type 72 AT
Source: People's Republic of China
Manufacturer: China North Industries Corporation, Beijing
Type: Anti-tank
Initiation: Pressure

This anti-tank mine, a copy of the Soviet TM-46, contains 5 kilograms of TNT and is regularly used by all three resistance factions. There are some indications that TM46 & 57 are also occasionally used.

Type 72
This multi-role anti-personnel mine is reportedly used by the Khmer Rouge, although there is no available information on whether the mines are captured from government forces or supplied by other sources (***see illustration above***).

Type 69
Source: People's Republic of China
Manufacturer: China North Industries, Beijing
Type: Anti-personnel/fragmentation
Initiation: Pressure or tripwire

This anti-personnel bounding mine is widely used by all factions in conjunction with other devices. The T69 can be set to explode by pressure or tripwire, and, on detonation, bounds to 1.5 meters before exploding and discharging approximately 250 fragments over a lethal radius of more than ten meters.

PMN-2

This is the most common mine in use in Cambodia. It is heavily deployed by government troops. Resistance forces, however, also refer to it as their own mine, calling it the "Singapore Mine." The Khmer Rouge are also reported to use the PMN-2 as a standard weapon. Resistance and government sources have said that gunfire is the only safe method of destruction. Gunfire is never a safe or sure method of eradication, but specialists agree that once armed, this mine should be destroyed.

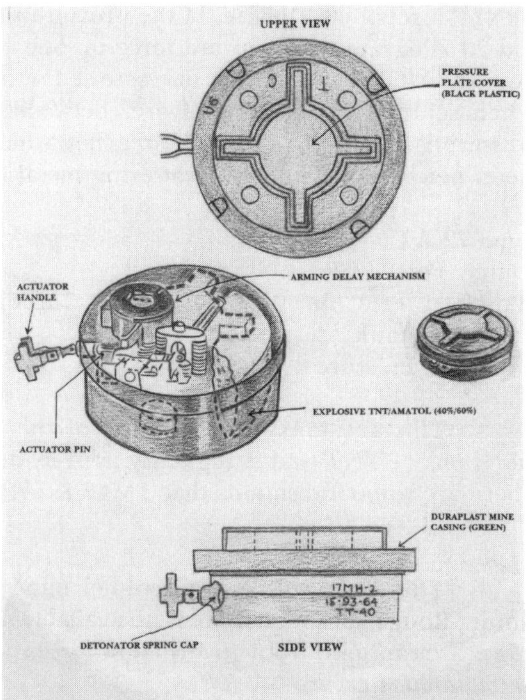

PMN-2

MD-82-B
Source: Probably Vietnam
Manufacturer: No data
Type: Anti-personnel blast
Initiation: Pressure

This anti-personnel pressure mine is used, though not frequently, by all three resistance factions. One was found laid in a border camp in Thailand in March 1991.

Valsella Valmara 69
Source: Italy/Singapore
Manufacturer: Italy - Valsella Meccanotecnica SpA, Brescia
Singapore - Chartered Industries
Type: Anti-personnel bounding fragmentation
Initiation: Pressure or tripwire

Descriptions of one mine used by resistance factions fit this Italian anti-personnel bounding mine, which is also manufactured by Chartered Industries of Singapore. When initiated, by pressure or tripwire, more than a thousand metal fragments are scattered over a lethal area of 25 meters at a height of approximately half a meter.

M-14
Source: USA (also produced in India)
Manufacturer: USA
India - Ordnance Factory Board, Calcutta
Type: Anti-personnel, non-metallic, blast
Initiation: Pressure

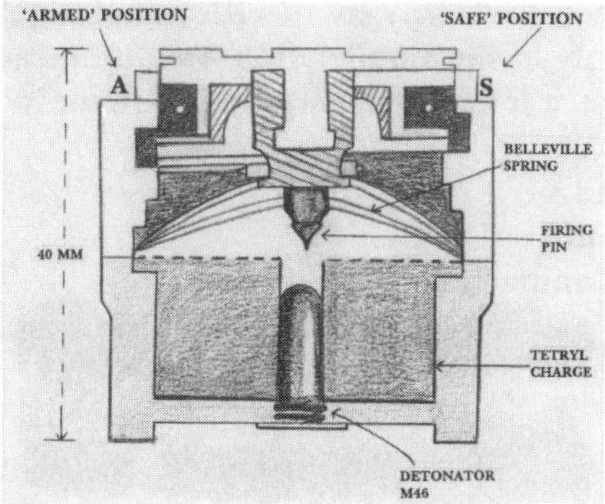

M14 ANTI-PERSONNEL BLAST MINE

This US-produced plastic pressure mine is used by the KPNLF and ANS forces. One informant identified it as a mine used by the Khmer Rouge, although this was not confirmed by other sources. The device is extremely compact, measuring only 56mm in diameter and 40mm in height, and weighs less than 100 grams.

PMD-6

Source: China, USSR and others
Manufacturer: (Easily produced locally)
Type: Anti-personnel blast
Initiation: Pressure

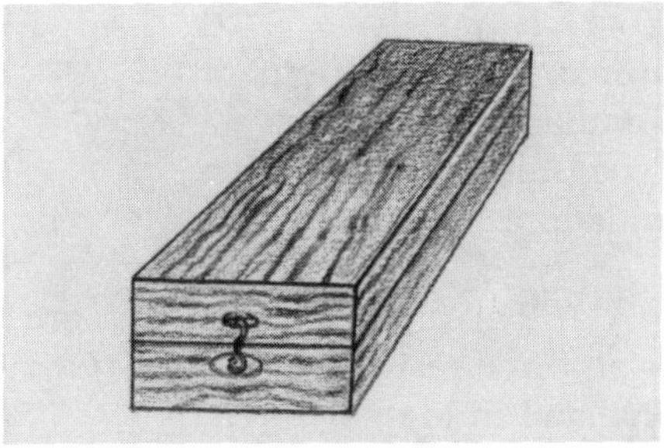

PDM 6

This simple but effective pressure mine consists of a hinged wooden box containing a block of TNT and a detonator. Variations of the mine can be locally manufactured without difficulty and may be tailored in size, initiation pressure, and explosive content to match a specific target.

Type-59
Source: China
Manufacturer: Unknown
Type: Anti-personnel fragmentation
Initiation: Tripwire

This mine is a Chinese-supplied version of the POMZ-2. It is used primarily by the Khmer Rouge. It differs from the Soviet version only in the smaller fragmentation case. Operated by tripwire, this mine has a lethal radius of approximately 20 meters.

M2A4
Source: USA
Manufacturer: Not known
Type: Anti-personnel bounding fragmentation
Initiation: Pressure or tripwire

This U.S.-produced device is a modified 60mm mortar bomb fitted with an M605 combination fuse to make a pressure or tripwire sensitive bounding mine. The bomb is projected between two and three meters high before exploding. The KPNLF use the M2.

M18A1 "Claymore"
Source: USA (Licensed & unlicensed production and copies by many countries)

Manufacturer: Morton Thiokol Inc. Shreveport, Louisiana
Type: Anti-personnel directional, fixed fragmentation
Initiation: Remote or tripwire

The Claymore is an American directional fragmentation mine manufactured at the Morton Thiokol/US Army facility in Louisiana. It is known to be deployed by the KPNLF. When exploded, usually by a pull wire or remote electric generator, 700 steel ball bearings are projected in a 60-degree arc for more than 50 meters to a height of six feet.

Improvised Explosive Devices and Booby-Traps

Resistance fighters, through foreign training, have learned the value of improvised explosive devices (IED's) and booby-traps in enemy-held territory.

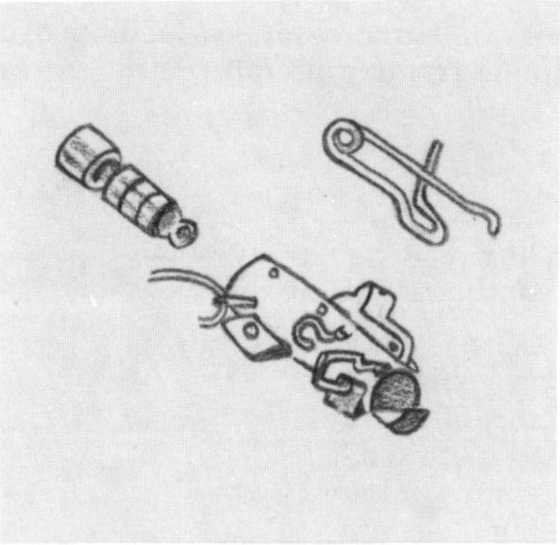

FIVE-WAY SWITCH

The **F1A1 Five-way Switch** illustrated above is commonly used by British, U.S., and other western armies. It is variously designated as **F1A1** and **M142**. The device's primary purpose is to afford an adaptable, easy-to-use method of improvising booby traps and is issued as a kit, complete with waterproof instruction sheet, which includes galvanized nails for securing the switch and a coil of filament tripwire. The KPNLF[60] use the F1A1 as standard issue and both ANS and Khmer Rouge fighters report its use as a standard tactic. Its

[60]KPNLF stocks of F1A1 Five-way Switch kept in base and forward stores give some indication of their reliance on improvised devices. In early 1991, they were being issued to forward areas in Cambodia at the rate of nearly 150 per month in conjunction with large quantities of TNT and CR plastic explosives.

most frequent reported use is in the trip, tension-release, or pressure role using an M79 mortar as the explosive charge. Some sources also reported use of the switch in conjunction with C4 plastic explosive.

Another initiation device, called the **Sheepshank Switch**[61], is a simple but effective improvisation used by resistance forces in conjunction with a battery and electric detonator to initiate the detonation of ordnance or plastic explosive. Resistance fighters told us that it is used in heavy undergrowth as a simple tripwire or, more commonly, attached to a door or attractive object. Moving the overlapping wires a critical distance in any direction will initiate the detonation cycle.

The **Mousetrap Electric Switch** is another home-made initiation device used by the resistance factions. It is a simple pressure switch, used in conjunction with a battery and electric detonator. Downward force causes the terminals to meet and complete an electric circuit and begin the detonation cycle. According to resistance fighters, it is used in Cambodia with a wide range of ordnance, most commonly mortar shells, as an explosive charge.

Three examples of improvised explosive devices from Khmer Rouge training literature are illustrated below:

Shows plan for a simple, but potentially very powerful and effective 'box-type' pressure mine using electric firing circuit.

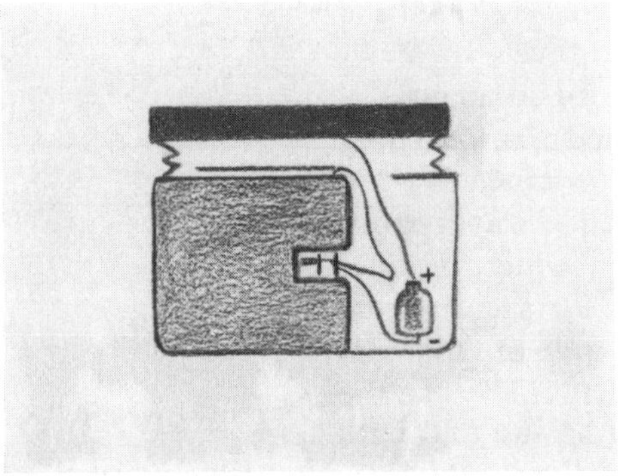

Plan for improvised "box-type"electric pressure mine:

[61] The name given here is a descriptive designation only.

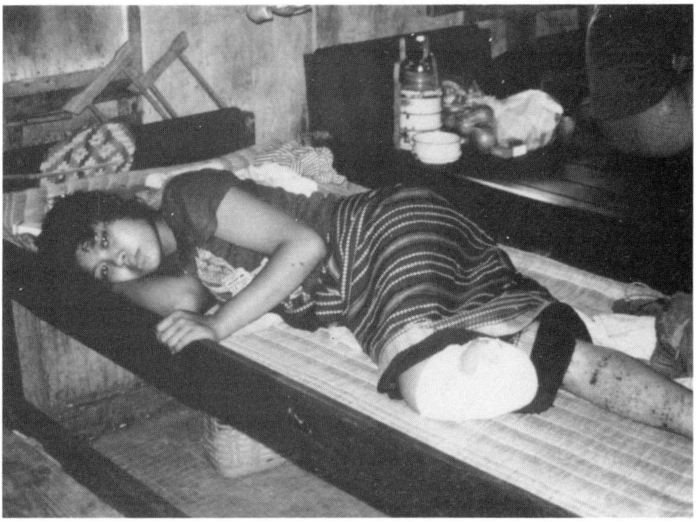

Nean Pok had only been married six weeks when she stepped on a land mine in April 1991.

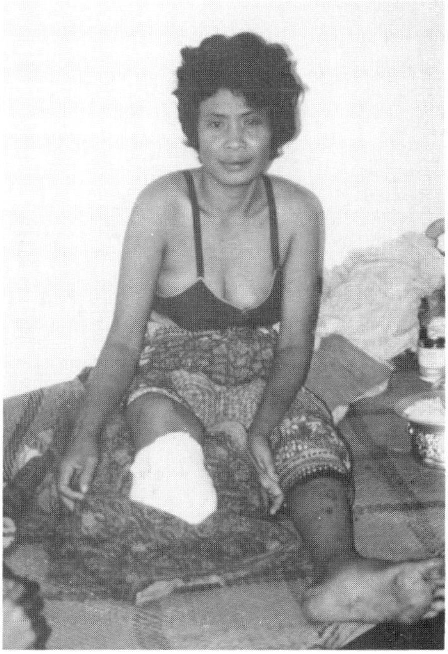

Ken Kop, a mother of seven children, stepped on a land mine while walking to work in the rice paddies.

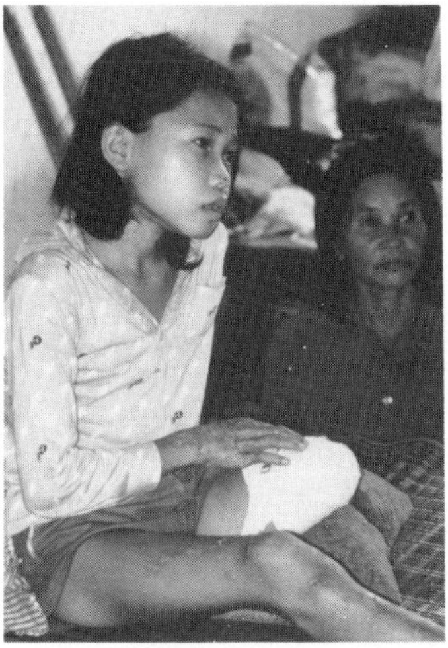

Six-year-old Chok Chuon lost her leg when she jumped on a mine while playing near a railroad line in April 1991.

Mine amputees flee a refugee camp in Thailand during shelling by Phnom Penh government troops.

Homemade protheses hang on a wall in a workshop run by Handicap International in Battambang.

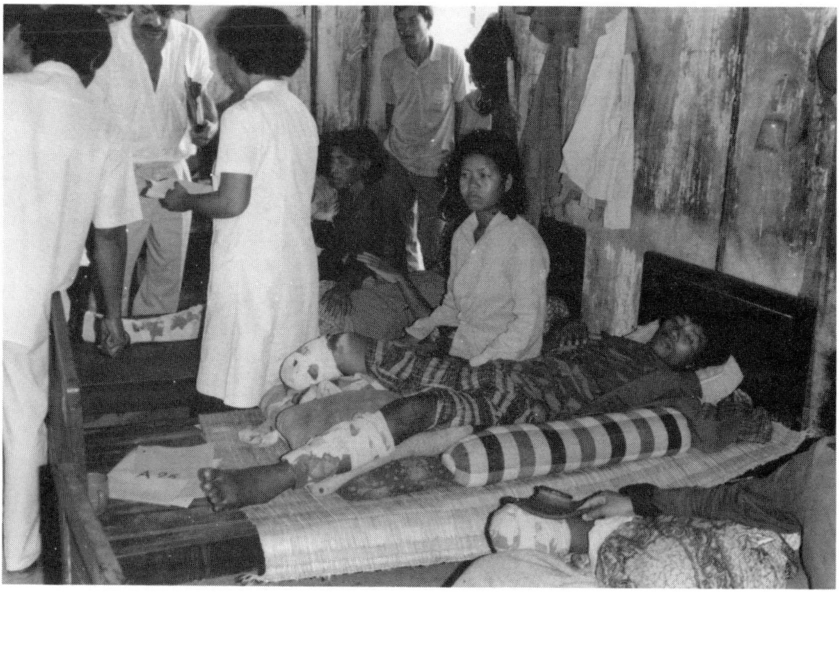

Physicians attend to mine victims in the Mongol Borei Hospital.

A sign in Site II, the largest of the Cambodian refugee camps in Thailand, warns camp residents of the danger of mines.

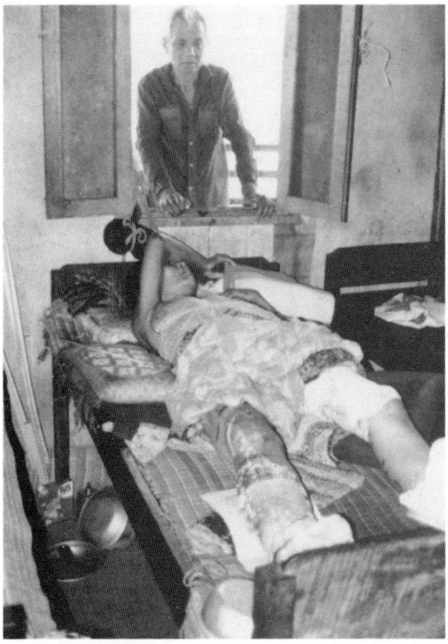

Praing Chhoeun stepped on a mine as she was taking her cattle out to graze. Her husband, standing in the window, was later forced to sell the animals to pay for medicines.

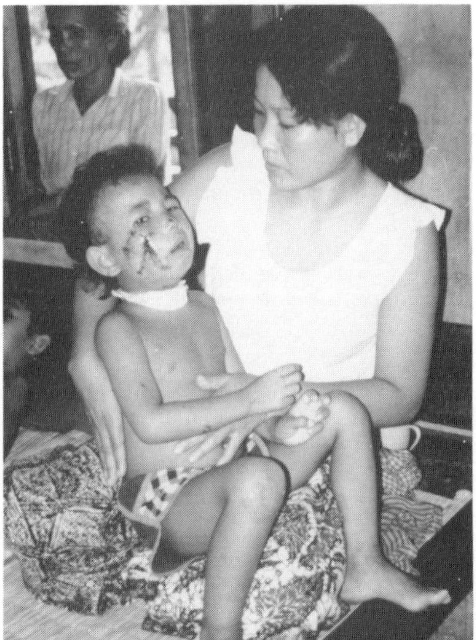

This child lost his mother and was badly wounded when KPNLF forces attacked Sala Kraw, a displaced persons camp in northwestern Cambodia, with rocket-propelled grenades in February 1991.

"Bow" tripwire booby-trap:
(Source: sketch from Chinese training manual for Khmer Rouge)

Improvised booby-trap using mousetrap and grenade:
(Source: Chinese Training Manual for Khmer Rouge)

Improvised electrically-initiated booby-trap using tripwire:
(Source: Chinese Training Manual for Khmer Rouge.)

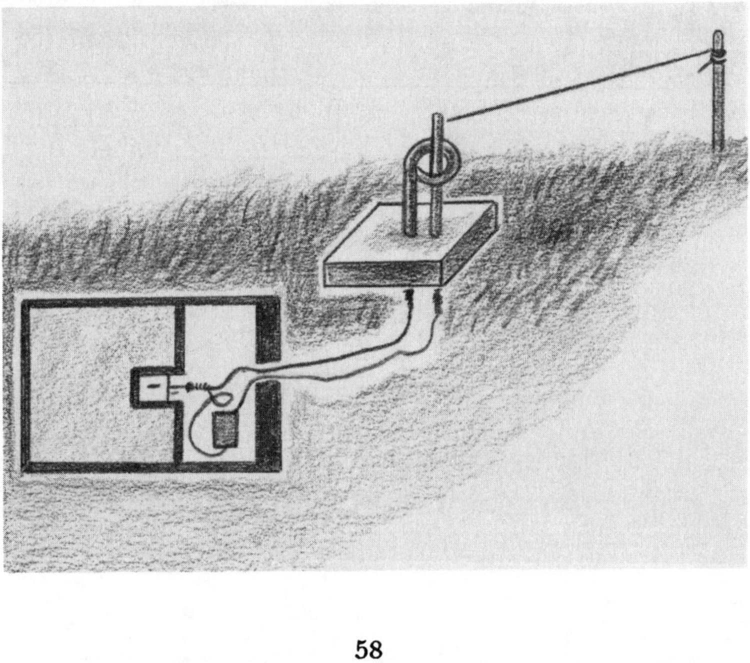

V. MEDICAL CARE

Cambodia today has the highest percentage of mine amputees of any country in the world. Surgeons in Cambodia perform between 300 and 700 amputations a month because of mine injures.[62] As a result, one out of every 236 Cambodians has lost one or more limbs after stepping on a land mine. By comparison, there are 60,000 amputees in Vietnam (out of a population of 75 million) who were crippled by the Vietnam War or by leftover debris such as unexploded mines, booby traps or artillery shells. This means that one out of every 1,250 Vietnamese is handicapped as a result of the war.[63]

According to Khmer and foreign surgeons working in Cambodia, for every mine victim who makes it to hospital another will die in the fields or on the way to hospital. No one, however,

---

[62]Surgeons in Cambodia perform between 300 and 700 amputations a month because of mine injuries. Given the duration of the war, the figure of 36,000 may actually be an underestimate.

[63]See D. McClellan, "New limbs for Viet amputees," *San Francisco Examiner*, May 12, 1991, p. A1. Angola's 28-year-old war produced perhaps 20,000 mine amputees (out of a population of 9.4 million). See Africa Watch, *Angola: Violations of the Laws of War by Both Sides*, April 1989. Official estimates put the number of war amputees in Mozambique at between 4,000 and 8,000 (out of a population of 14.9 million). In Uganda where war, infections, and improper medical care have created a high number of disabled, there are 15,000 amputees (out of a population of 16.6 million). See "Planning for Improved Orthopaedic and Prosthetic-Orthotic Programs in Uganda and Mozambique," a report prepared of the U.S. Agency for International Development, U.S. Department of Public Health and Human Services, June 26, 1989.

knows exactly how many have died, nor is it likely that anyone ever will. No institution has kept records of war-related deaths among civilians. Moreover, Cambodians, being mostly Buddhist, burn their dead.

What has become increasingly clear, however, is that land mines have injured (and possibly killed) more combatants and noncombatants alike than any other weapon in Cambodia's 12-year-old civil war. In Khao I-Dang Hospital, the largest hospital for Cambodian refugees in the Thai border camps, 57 percent of those treated for war wounds from January 1990 through March 1991 had been injured by land mines.

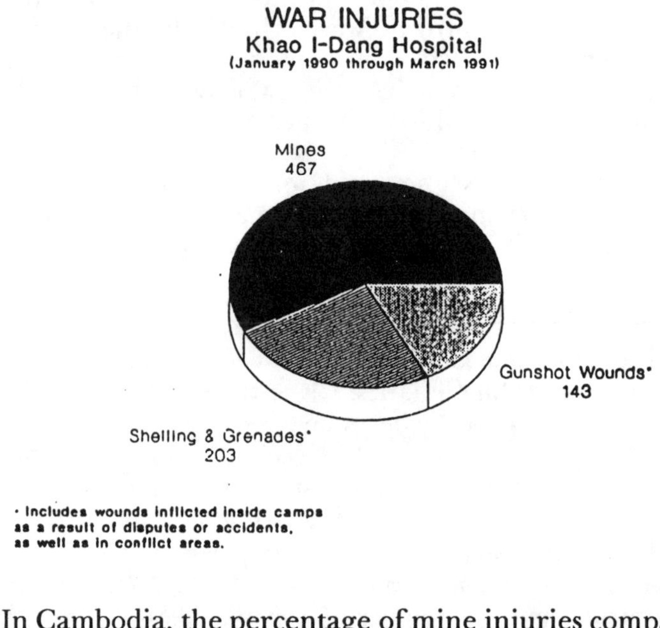

In Cambodia, the percentage of mine injuries compared to other war-related injuries was the same or slightly higher. Dr. Chuon Bunthorl, the director of the 329-bed provincial hospital in Battambang, told us that just over 50 percent of the war wounded who arrived at the hospital in 1990 were land mine victims. In a civilian hospital near Siem Reap, of the 60 patients with war-related injuries, 45 of them were injured by mines.

A visitor to Cambodia cannot help but be struck by the number of amputees; they are everywhere. For instance, in only a few hours travelling along Route 5, the war-torn strip of highway that connects Phnom Penh with the northwest, we saw a legless barber in a wheelchair shaving a customer under a tarp he had fashioned into a tent. Then, further on, a five-year-old boy, his crutch tucked under his right arm, stood by a bomb-blasted bridge. As cars slowed down, he thrust his cap out, angrily demanding money. In the Battambang market, three women amputees sat together on the remnants of an old cardboard box, selling fruits and vegetables.

Travelling to hospitals and prosthetic workshops in Cambodia and in the Thai border camps, our delegation interviewed Cambodians, both combatants and noncombatants, who had been injured by land mines, as well as Khmer and foreign doctors, to document the prevalence and types of mine injuries. Most importantly, we wanted to know what happened to civilian mine victims from the moment they stepped on a mine until their discharge from hospital. Using a set of pre-determined questions, we asked them to describe their ordeal to us. Among the questions we posed to them were the following: What were they doing at the time they encountered the mine? Had others in their family or village been killed or injured by mines? What sort of first aid, if any, did they receive immediately after the blast? How much time passed from the point of injury to their arrival at the hospital? Did they receive anaesthetic during surgery? We also asked mine victims still in hospital about their prospects for receiving a prosthesis and physical therapy and being reintegrated into society.

Our observations are based on their responses and interviews with medical and relief personnel--some of whom have worked with Cambodian mine victims for the past six years.[64] We also gathered data on war injuries, including mine-related

[64]In addition to Cambodian medical personnel, we interviewed physicians working with the International Committee of the Red Cross, the Swiss Red Cross, and Médecins du Monde.

wounds, and amputations from six hospitals[65] inside Cambodia and Khao I-Dang Hospital, the largest surgical hospital in the Thai border camps. We were unable to collect data from every hospital in Cambodia either because of limited access to hospital records or because records had never been kept. However, the hospitals that provided data were, by and large, those which were in or close to combat zones and thus treated a large number of mine injuries. We believe this data, coupled with our interviews, provides an accurate picture of the scope and nature of mine deaths and injuries in Cambodia.

The Health Care System

To understand the quality of medical care available to Cambodian land mine victims, it is first necessary to take a brief look at the history and development of Cambodia's health care system since the Khmer Rouge took power in April 1975. Within days of their arrival in Phnom Penh, the Khmer Rouge expelled the ICRC from the country and closed its borders to all foreign medical agencies. Over the next three and a half years, the Khmer Rouge destroyed Cambodia's entire health care system--equipment, supplies, and buildings.[66] In its place, they constructed small regional health clinics, and staffed them with ill-trained cadres from within their ranks who were selected by political rather than medical criteria. These "health workers" scorned modern medicine and instead practiced traditional medicine, but their understanding was based more on superstition and folklore than a genuine knowledge of herbal remedies.

By early 1979, when the Vietnamese ousted the Khmer Rouge from power and installed a puppet government in Phnom Penh, Cambodia had one of the most wretched health care systems in the world. Of the 450 medical doctors in Cambodia before

---

[65]The six hospitals were located in Kampot, Takeo, Mongol Borei, Battambang, and Kampong Chhnang.

[66]See L. McGrew, "Health Care in Cambodia," *Cultural Survival* 14, no. 3:77.

1975, only 45 remained, and of those, 20 left the country after the Vietnamese invasion.[67] Large sections of the population were suffering from tuberculosis, malaria, ankylostomiasis, respiratory, and infectious diseases--all of which were compounded by malnutrition. There were virtually no nurses and a severe shortage of medicines.

Since 1979, in spite of the civil war and a paucity of international aid, the Phnom Penh government has made some progress in reconstructing and rehabilitating the health delivery system in Cambodia. But this development has only taken place in recent years. In 1983, for instance, a team sent to Cambodia by the Food and Agriculture Organization found that the health system was "disastrous." Medical supplies were "far below acceptable standards, even for poor developing countries, and the situation is nearing a deep crisis....Adequate medical treatment is not available because of a general lack of basic medicines, a severe shortage of medical doctors, and absence of basic medical supply and coordination."[68]

Although many of these deficiencies were remnants of the Khmer Rouge era, they also reflected Vietnamese policies that gave politics precedence over health. "Those few Cambodian doctors and nurses who did exist were constantly forced to neglect their duties to go to political study sessions," wrote British journalist William Shawcross.[69] "There was daily indoctrination, and there were frequent longer courses. Patients died as a result." The Phnom Penh government welcomed aid agencies willing to build more clinics and hospitals, useful propaganda for the regime, but were less enthusiastic about non-Communist health professionals training local medical personnel.

---

[67]See E. Mysliwiek, p. 42.

[68]FAO report is quoted in W. Shawcross, *Quality of Mercy*, p. 399.

[69]See W. Shawcross, p. 400.

As of mid-1991, Cambodia had 10,000 hospital beds--a ten-fold increase since the Khmer Rouge period--and more than 500 Khmer physicians, as well as over 7,000 nurses, medical assistants, and midwives. However, international health assessment teams have routinely found them to be poorly paid, poorly trained, and isolated from the mainstream of public health and primary care developments in other parts of the world. Given the poor quality of healthcare, the impact of mines is all the more appalling -- if returning refugees get hurt, what facilities will be available to treat them?

Government health officials point to the national expanded program on immunization, started in 1986, as one of their most notable achievements in health care. The program has fully vaccinated 60 to 70 percent of children in Phnom Penh and 12 of the country's 19 provinces.[70] There are wide gaps, however, in the program's coverage because medical personnel cannot enter certain conflict zones.

For most Cambodians, access to health care is a luxury. In 1990, only 53 percent of the population had access to health services.[71] Often those in greatest need live in rural areas where living conditions foster illness and disease. Cambodia's infant mortality rate (IMR) now stands at 133 per thousand, placing it in UNICEF's worst category, and, in many parts of the country, the physician/patient ratio may reach as high as one per 30,000 or

---

[70]See N. Hirschhorn, L. Haviland, and J. Salvo, "Critical Needs Assessment in Cambodia: The Humanitarian Issues," a report to the U.S. Agency for International Development, April 1991, p. 6.

[71]See United Nations Development Program, *Human Development Report 1991* (Oxford: Oxford University Press, 1991), p. 123. The UNDP ranked Cambodia 140 out of 160 countries for 1990 based on its human development index, a set of indices used to measure such things as health delivery, income distribution, life expectancy at birth, and education.

more.[72] Fifteen percent of tested blood carries hepatitis. Epidemic diseases, such as malaria, tuberculosis, diarrhea, and dengue fever, are widespread, and, according to an AID assessment team that visited Cambodia in April 1991, "good evidence exists that preconditions for [an] upsurge of these diseases are present."

Today there are several foreign medical relief organizations operating in Cambodia.[73] For the most part, these groups confine their activities to a particular hospital or health problem. Since 1981, the French Red Cross has supported a national tuberculosis control program. World Vision maintains a large children's hospital in Phnom Penh. And across the city, at the Calemete Hospital, doctors with Médecins du Monde are constructing a new surgical unit.

Since 1979, the ICRC has set up surgical teams in Kampot, Pursat, and Mongol Borei primarily to treat the war injured. Because of the large number of war wounded, the ICRC have begun first-aid training programs at local health clinics. In October 1990, the ICRC, in conjunction with the Ministry of Health, revived the National Blood Transfusion Center in Phnom Penh. Early in 1991 the blood bank's mobile units began collecting blood in various parts of the city by offering donors a meal and a Red Cross T-shirt.

Such progress notwithstanding, our delegation found that blood was in critically short supply in nearly every hospital we visited in Cambodia. For land mine victims, the availability of blood can often mean life or death. We also found that hospitals in or near conflict areas were in serious need of X-ray film, anesthetics, antibiotics, latex gloves, and surgical supplies.

---

[72]By comparison, in many parts of Sudan, which along with Cambodia ranks as one of the world's most impoverished countries, the physician/patient ratio may reach as high as one per 24,000.

[73]Among them are Médecins du Monde, World Vision, Médecins Sans Frontieres, and several teams from Red Cross societies.

Electricity is also in short supply in hospitals throughout Cambodia. In provincial hospitals, electrical power is often available only in the evening from 6:30 to 10:30 p.m. During emergency treatment, backup generators are used to run the operating theaters. However, they frequently break down or cannot be operated because of the lack of diesel.

Many of Cambodia's provincial hospitals do not have indoor plumbing in the wards, and water must be brought to the hospital in five-gallon jerry cans. The plumbing in operating and treatment rooms has deteriorated. Patients who are ambulatory must bathe and drink from outdoor pumps. The facilities for sterilization, disinfection, and surgical instruments are very limited, resulting in many infections after surgery.

Mine Injuries

Cambodian hospitals are poorly equipped to deal with war injuries. The Mongol Borei hospital, located just south of Sisophon, was so crowded with patients when we visited in April 1991 that many of them were sleeping outside on cots. Most of the patients in the surgical wards were victims of mine blasts. They lay on bamboo mats or propped themselves up against soot-black walls, as flies swarmed about their bandaged stumps. Many suffered from chronic anemia or malaria.

Our host, Dr. Chris Giannou, a Canadian surgeon with the ICRC, took us through the surgical wards. Dr. Giannou had spent most of the 1980s as a war surgeon and hospital administrator in war-torn Lebanon. "In Lebanon you became an expert in the cacophony of war," he explained. "You could distinguish the sound of incoming and outgoing artillery--whether it was an 80- or 160-millimeter mortar, or a Howitzer, and so on--and the sound of the damage it made. So, as a surgeon, I treated lots of shell and bullet wounds because mines weren't such a problem there. But here it's different..."

Moving from cot to cot, he stopped next to the bed of a woman named Praing Chhoeun. Three days earlier, she had stepped on a mine while herding her cows and spent 12 hours in an oxcart and on the flatbed of a truck before she made it to the hospital. Dr. Giannou carefully lifted back her sarong and revealed a badly infected stump. "Now if that had been caused by shrapnel from an artillery shell, you would do a simple debridement, clean it up, no problem," he said. "But these mines drive dirt and bacteria as well as the shrapnel up into the tissue. So infection spreads fast. Then there is the effect of the shock wave, which causes blood vessels to coagulate and thrombose well up the leg. So I end up having to amputate much higher up than where the wound appears."[74]

The work of Dr. Giannou and other surgeons in Cambodia is further complicated by the use of plastic shrapnel or casing on land mines. Once embedded in tissue or bone, these dark, frog-green plastic fragments, unlike metal fragments, are difficult to detect on X-rays, and therefore must be located visually and then extracted.[75] However, they are often overlooked during the surgical removal of foreign matter and dead tissue from the wound. And if the fragments are not removed, they can later cause serious infections, including osteomyelitis, an infection of the bone cortex and marrow.

Protocol I of the 1981 U.N. Convention, known as the "Protocol on NonDetectable Fragments," states: "It is prohibited to use any weapon the primary effect of which is to injure by fragments which in the human body escape detection by X-rays."[76] Do all four parties in the Cambodian conflict intentionally use mines containing plastic shrapnel or plastic casing because they know that these elements cannot be detected

---

[74]See also R. Fasol, S. Irvine, and P. Zilla, "Vascular injuries caused by anti-personnel mines," *Journal of Cardiovascular Surgery* 30: (1989):467-472.

[75]Descriptions of the shrapnel given by Dr. Giannou and others suggests that the PMN-2 mine is the major source of such secondary infections.

[76]See A. Roberts and R. Guelff (eds.), *Documents on the Laws of War* (Oxford: Clarendon Press, 1982), p. 475.

on radiographs? Or do they prefer to use mines constructed with plastic because they are lighter and thus easier to carry? Or do they simply accept whatever mines are sold or supplied to them? Whatever the answer, all four warring parties have violated if not the letter than the spirit of the protocol, as have the manufacturers and supplies of these mines.

Dr. Giannou and other surgeons in Cambodia and in the border camps in Thailand have found that if amputation is necessary when treating lower-body mine injuries, it is usually required below the knee. In 1990, for instance, 63 percent of all lower-limb amputations performed on patients injured by mines were below the knee in hospitals in Kompot, Pursat, Takeo, and Khao I-Dang. Some victims of mine blasts suffer injuries above the waist. Fishermen who snare small anti-personnel mines in their nets frequently suffer upper limb and facial injuries. Similarly, combatants and non-combatants alike frequently suffer upper body injuries when they attempt to defuse or move mines. Peasants are most likely to step on mines when they are looking for firewood, herding animals, working in the fields, fishing, or simply walking to another village. The internally displaced are particular at risk when they venture out of their camps to forage for food and firewood.

By April 1991, when we visited Cambodia, 186,000 villagers had been displaced in 9 provinces as a result of the war. Most of them were living in camps in the northwestern part of the country where the fighting had been most severe. During our visit to the ICRC hospital in the village of Mongo Borei, we interviewed several patients who had stepped on mines near camps for displaced persons. Doctors there also told us that three boys had recently been killed when they returned to their village from a displaced persons camp. According to the doctors, at midnight on February 19, about 150 KPNLF soldiers entered the Sala Kraw camp, located 8 kilometers north of the provincial capital of Sisophon. Before storming the camp, soldiers launched rocket-propelled grenades into the camp, killing nine civilians, including two children aged 4 and 10, a pregnant woman, and a

75-year-old man, and wounding 15 others.[77] Inside the camp the soldiers reportedly destroyed a recently constructed school house, torched hay and rice stores, and stole several motorcycles.

Days after the KPNLF raid, camp residents, terrified that their attackers would return, sent a young boy back to their village three miles away to see if it was safe to return. But he stepped on a mine and was killed. When residents learned of the boy's fate, they dispatched two older boys to the village. They, too, were killed by land mines.

For the past four years, Dr. Johannes Schraknepper, a surgeon with the Swiss Red Cross in Takeo province, has attended to thousands of victims of mine blasts. In his experience, soldiers who are wounded by mines usually have better access to transport and thus arrive at hospital much sooner than civilians. He characterized the situation for noncombatants as follows:

> For civilians, finding transportation is a big problem. If they're lucky, someone--most likely a relative or friend--will find them wounded in the field and will apply a tourniquet. It stops the bleeding, which is good, but too often they forget to loosen it, which causes problems later. So the wounded person will lie in the fields, or maybe in his house, while someone goes looking for transport, which is usually a horse or motorcycle taxi. Now, first they have to find it, which isn't always easy, as it may require riding a bicycle five miles away to another village. In the meantime, the family had better have enough money, because usually the rule is no cash, no transport. All of this may take 6 to 12 hours or even an entire day. Then it may take several more hours to get to the hospital.

---

[77]The ICRC promptly protested the attack in a letter from its delegation chief in Cambodia, Jean-Jacques Frésard, to KPNLF president Son Sann, dated February 21, 1991.

Once the injured reach hospital, they often find that there is little or no food, except what relatives bring. In some hospitals, if meals are provided, war veterans are given first priority.[78] Even though health care in Cambodia is supposed to be free, Cambodian doctors and nurses regularly charge their patients for services, medicines, and intravenous fluids. If blood is needed, the patient's family must find donors and pay them. Lach Pem, a 55-year-old farmer who stepped on a land mine in 1987 and eventually fled to a refugee camp in Cambodia, said that he wasn't sure which was worse: losing a leg or knowing that his wife had gone to relatives and friends to beg for money to pay for his hospital care.

In mid-1991, a Cambodian doctor earned about $13 a month, and a nurse or laboratory technician earned about $7 a month. Yet, it takes about $40 a month just to survive. As a result, many Cambodian health professionals spend as little time as possible at their official posts and either work in private practice or run a business, such as a pharmacy, to sustain their families.

The following accounts are drawn from our interviews in April 1991 with civilians who were injured by land mines:

■ **Nean Pok**, a 20-year-old woman receiving care in the Mongol Borei hospital, told us of her ordeal, as her husband listened, stopping her from time to time, to add details to the account. They had been married six weeks earlier. On April 6, 1991, at about 12 p.m., Nean Pok stepped on a mine, possibly a PMN-2, while gathering firewood at the edge of the forest near Phrum Prey Kpors, a village 30 kilometers southwest of Mongol Borei. Hearing the explosion, her husband rushed to the scene. He fitted a tourniquet around her left leg, carried her to the side of the road, and flagged down a moto-cyclo, or motorcycle taxi. He took her first to a local first-aid post and, seven hours later, to the hospital where surgeons amputated her lower left leg. Nean Pok and her husband were able to identify several mines from

[78]See N. Hirschhorn et al, "Critical Needs Assessment in Cambodia: The Humanitarian Issues," p. 8.

photographs. There were no mine markers near the village, and mines had killed or injured several villagers and farm animals.

■ **Chang Song** is a 38-year-old fisherman from the village of Phrum Chek. He is married, has four children, and regularly fishes on a one-kilometer long lake near his home. Early in the morning on March 30, 1991, Chang Song and several other fishermen had spread themselves out, at ten-meter intervals, along the lake's shoreline. Just as he was about to toss his net into the water, he heard a loud explosion and remembers collapsing to the ground in pain. Other fishermen picked him up by his clothes and carried him back to his home. Four hours later, he was taken to a local first-aid post and fitted with a tourniquet. At 5 p.m. that evening, he arrived at Mongol Borei hospital. Because his wounds were so heavily infected, surgeons chose to amputate his right leg above the knee.

■ On April 6, 1991, at approximately 3 p.m., **Ken Kop**, a 42-year-old mother of seven children, was following her usual riverside route to work in the rice paddies when she stepped on a mine. Her brother carried her in his arms to her house, where he and other relatives hastily made a sling out of a hammock and bamboo pole and rushed her to the district hospital. Ken Kop's right leg was so badly mutilated, that the attending doctor immediately amputated it above the knee. The operation was carried out without anaesthesia. Later that evening, she arrived at the provincial hospital in Battambang, where the wound was closed under general anaesthesia.

■ At sunset on April 11, 1991, **Praing Chhoeun**, a 56-year-old farmer, stepped on a mine as she was taking her cattle out to graze for the night. It was a trip she made every morning and evening. After the explosion, Praing Choeun apparently went into shock and recalled very little of what took place the rest of the day. Her husband, who had been with her in hospital since the accident, told us that he took his injured wife by ox cart to the Sosphean district infirmary. The following day, he arranged for a truck to take them to Mongol Borei hospital. When asked if the countryside around their village was mined, Praing Choeun nodded and then added that no one knew exactly where they were buried. "I had always worried about stepping on one," she said, "but then the cattle had to be grazed."

■ Six-year-old **Chok Chuon** lost her left leg when she jumped on a mine while playing near a railway line on the morning of April 6, 1991. Her mother heard the explosion and rushed to the railroad tracks and carried her home. According to Chok Chuon's mother, there were no markers warning of the presence of mines. Another relative fixed a tourniquet to Chok Chuon's left leg and then, with the help of others, carried her in a sling to the main road, 15 kilometers away, where they flagged down a moto-cyclo. At 2:30 p.m., she arrived at the provincial hospital in Battambang and, two hours later, went into surgery, where doctors performed an above-knee amputation.

■ Fifty-five-year-old **Lach Pem** and his wife are "displaced" Cambodians who arrived at the Site II border camp in March 1991. They are originally farmers from the Moung District in southern Battambang province. Years earlier, on September 9, 1987, Lach Pem stepped on a land mine while gathering firewood in the forest. (In 1984, his eldest son, Chhim Pang, had stepped on a mine while fighting with the KPNLF.) Five of Lach Pen's friends carried him in a hammock-sling to Moung District Hospital. The trip, on foot, took 20 hours. After his amputation, he developed a serious infection and had to remain in hospital for three months. During that time, he spent 15,000 riels, or about $150, on medications. After leaving the hospital, he bought crutches and returned to work in the rice paddies. Soon after his arrival at Site II, Lach Pen learned that during his journey to Thailand, another son, a noncombatant like his father, had stepped on a mine and was in the Moung District hospital.

Social and Psychological Aspects

Cambodia is an agrarian society where muscle power means survival.[79] Nearly every aspect of a Cambodian's life is set to the rhythm of rice cultivation--the flooding, the planting, the replanting, and harvesting. It is very labor intensive, requiring the participation of every man, woman, and child. And a person who is physically disabled can become a burden--someone who eats but produces nothing.

Most amputees leave hospital with little hope for the future. There are no rehabilitation centers, and Cambodia has no laws to protect amputees against discrimination or exploitation. Female amputees are less desirable as wives because they cannot work in the fields, and male amputees are not allowed to become Buddhist monks. Many amputees drift to Phnom Penh or larger towns and become beggars or petty criminals.

In most peasant cultures, the village and extended family are almost synonymous, nurturing a solidarity that sustains both the individual and his or her community through difficult times. In Cambodia, however, 20 years of famine, genocide, foreign occupation, and civil war have undermined the communalism that once existed. In Mongol Borei hospital, Dr. Chris Giannou recalled the plight of a young child who had become paraplegic after stepping on a mine:

> At first, the family didn't know what to do. So they abandoned him at the hospital because there was nothing left to expect from him. He stayed alone at the hospital for four months before they finally came back and got him. But the mere fact that they

---

[79]Eighty-eight percent of Cambodia's population lives in rural areas, 3 percent of which are involved in agricultural production. In the developing world, only five other countries--Bhutan, Burundi, Burkina Faso, Nepal, and Oman--have a larger percentage of rural inhabitants. See United Nations Development Program, *Human Development Report 1991*, p. 13-137.

> abandoned him...and we didn't know if they were going to come back or not. And they probably didn't know! In a Third World society, in a peasant culture, that's a sacrilege, it's unthinkable...only in Cambodia.[80]

Amputees often find that they cannot compete with the able-bodied for farm land, even though they can still supervise the farming or actually till the fields themselves. In 1988, the Phnom Penh government formally abandoned its policy of collective farms and began a program of land reform. Land was divided based on the number of active adults in the family. As a result, families with amputees received less land or less valuable land than families without amputees.

According to Handicap International (HI), the Belgian-based organization that runs 13 prosthetic workshops throughout Cambodia, Khmers often do not hire amputees, even after they've received training in a particular skill. Maite Idiart, HI director in Phnom Penh, estimated that only 20 percent of Cambodia's amputee population will find work.[81] The Jesuit Refugee Service (JRS) has started a small program to counteract this problem in a village near Phnom Penh. If the project is successful, there are plans to extend it to other villages that have a high percentage of amputees. Small no-interest loans are given to poor families to help them expand an existing business or start a new one. In turn, the families are contractually obligated to hire a disabled person who will be trained by JRS.

During our visit to Thailand, we met with Abbot Mony Chenda at the Buddhist temple in Site II, the largest of the Thai border camps. In addition to his role as a religious leader in the camp, Abbot Mony runs a pottery workshop for abandoned children and the physically disabled, many of whom are victims of mine blasts. We asked the abbot if it was true that young boys

---

[80]Interview, Mongol Borei, April 13, 1991.

[81]Interview, Phnom Penh, April 9, 1991.

(only men can become Buddhist monks) who were amputees could not become a bonze, or monk. "It is a rule that a bonze who is ill cannot go more than three days without collecting alms," he said. "So a boy or man who is amputated would be a burden to his fellow monks." He went on to say that Buddhist teaching in Cambodia emphasized inner and outer "wholeness," so an amputee could not be ordained as a monk.[82]

Cambodian amputees often appear stoic. In interviews, they speak frankly and deliberately about the events leading up to the moment when they stepped on the mine and their subsequent journey to hospital. But to give details, especially to a Westerner, about how they are "coping" mentally with their trauma would seem inappropriate. Some may feel that they are somehow responsible for their suffering because of the Buddhist concept of "karma."[83] A Swiss surgeon remarked that he had seen amputees, particularly young women who had been horribly disfigured, put "a mask over their despair" in the company of relatives and friends. Several refugee workers reported that Khmer amputees had a high incidence of alcoholism and suicide, which is unusual in Buddhist culture.[84] But, to date, no one has examined these problems and compared their incidence to the population as a whole.

---

[82]Interview, Site II camp, Thailand, April 17, 1991. Such religious attitudes toward the physically disabled are not confined to Theravada Buddhism; they can also be found in many Christian religions.

[83]This same response has been seen among Cambodians refugees living in the United States who either experienced torture or witnessed atrocities under the Khmer Rouge. See R.F. Mollica, G. Wyshak, and J. Lavelle, "The psychological impact of war trauma and torture on southeast Asian refugees," *American Journal of Psychiatry* 144 (1987):1567-1572.

[84]See, for example, P. Carey, "Cambodia's Unending Agony," p. 32, and N. Hirschhorn, L. Havviland, and J. Salvo, "Critical Needs Assessment in Cambodia: The Humanitarian Issues," p. 36.

Treatment and Rehabilitation

Cambodia has a very limited institutional ability to deal with the disabled. Children with polio, congenital learning or emotional disorders, or disabilities like blindness cannot be cared for, and many of them are abandoned at hospitals and orphanages. By its own laws, the Phnom Penh government is supposed to provide the disabled with a monthly pension, but relief agencies report that these payments are often paid in a single lump sum soon after the accident or never paid at all.

The Ministry of Social Action is responsible for the care of the disabled in Cambodia. But because of budgetary restraints, the lion's share of the work is done by administrators and technicians with Handicap International (HI) and the American Friends Service Committee (AFSC). In 1982, the two groups began to support Cambodia's first prosthetic workshop in Phnom Penh, where amputees are fitted with artificial limbs. Since then, they have established 12 more workshops throughout the country, where expatriates fit prostheses and train Khmer technicians. HI has also launched a two-year training program for nursing students in physical therapy at the nursing college in Phnom Penh.

Despite these valiant efforts, the number of artificial limbs they turn out (1,300 a year) falls far short of the demand. Only one in eight amputees receives an artificial limb, and most of them are soldiers. At the present rate it will take over 25 years to handle the existing waiting list of mine victims. By law, the Ministry of Social Action is required to fit soldiers with artificial limbs before civilians. After discharge from hospital, soldiers, especially higher ranking officers, are often transported individually or in groups to the prosthetic workshops. Amputees in the military hospital in Sosophon, for example, are taken 70 kilometers to the prosthetic workshop in Battambang, or even as far away as Phnom Penh.

Civilians, on the other hand, are discharged from hospital and left to fend for themselves. Even though they may have heard about the workshops from fellow patients or hospital staff, they often opt to return directly to their villages rather than pay the extra expense. Many are not even aware of the advantages of prosthetic devices. Back home, a few will fashion limbs out of scraps of wood and metal. (One amputee in a displaced person camp near Sisophon even converted an old artillery shell casing into a prosthesis.) But most amputees will grow accustomed to their crutches and, as time passes, simply keep postponing the trip to the workshop for financial or other reasons.

In Cambodia, prosthetic devices, like health care, are supposed to be free. But with inflation (in 1988 one dollar brought 150 riels; in 1991, 650), some Khmer technicians have begun supplementing their salaries (4,000 riels a month) by charging patients for artificial limbs. Although HI/AFSC disapprove of this practice, they realize that a "fees-for-service" attitude is universal in Cambodia and if they tried to prevent it in their workshops some of their best technicians would leave at a time when they are struggling to meet demand.

By using local materials (wood, leather, and locally processed rubber), HI/AFSC avoids the expensive fittings and complex moving parts required for more sophisticated artificial limbs. HI/AFSC can produce a below-knee prosthetic device for $12, and an above-knee device for $20. They can also open a workshop for less than $1,000. Unlike other prosthetic techniques, the production of their artificial limbs requires no power, crucial in a country like Cambodia where both electricity and diesel fuel for generators are in short supply. In contrast, a highly sophisticated, computer-based system in Vietnam, called the "Seattle Foot" by its American inventor, can produce a prosthetic device for $190, but it is virtually useless without power.

The HI/AFSC system, however, has its drawbacks. Leather, which is used for the knee socket, gradually loses its shape as it gets wet, and must be replaced every two years. In the past year, HI/AFSC has found quality leather more difficult to obtain as Cambodian leather wholesalers, encouraged by the government's relaxation on export controls, have begun quietly selling their best-quality leather to Thai businesses. Some amputees complain that the device's rubber foot breaks within a year and that the rigid wooden limbs, particularly on the above-knee devices, are cumbersome and unsuitable for work in rice paddies.

By the end of 1991, the ICRC plans to supplement HI/AFSC's work by opening orthopaedic workshops in Phnom Penh and Battambang.[85] ICRC's goal is to produce 1,000 prostheses per month for all of Cambodia. The new legs will be lighter, stronger, more flexible, and more comfortable than the traditional models. The ICRC also plans to work independently of the Ministry of Social Action by hiring local technicians and paying them at ICRC-established rates. This should increase productivity dramatically but could potentially make the entire prosthetic industry dependent on outside funding.

Cambodian amputees in the Thai border camps appear to fare better than their counterparts at home. According to HI, which operates workshops in eight camps, nearly every amputee who wants a prosthetic device will eventually get one. Since 1984, HI has trained over 70 Khmers in the camps as prosthetic technicians and physical therapists. Another relief organization, the Catholic Office for Emergency Relief and Refugees (COERR), runs three vocational training schools for camp amputees who can learn watch repair, carpentry, typing, radio and TV repair, welding, and engine repair. The logic behind offering these skills is based on the presumption that after repatriation to Cambodian, many, if not most, amputees will eventually drift to the cities in search of work.

According to Ky Ka, the Khmer director of the COERR training school in Site II, amputees in the camps may be better off there than in Cambodia but they have also suffered from years of dependency and boredom.[86] As a result, he says, a disproportionate number of mine victims have become petty criminals and alcoholics. Many amputees become despondent and

---

[85]The Cambodia Trust, an Oxford-based charitable organization, also plans to open a prosthetic clinic in Calamete Hospital in Phnom Penh. In addition, the Vietnam Veterans of America Foundation is finalizing a program that would help provide prosthetics for 79 injured Cambodian veterans living in a camp across the Tonle Sap River from Phnom Penh.

[86]Interview, Site II, Thailand, April 17, 1991.

frequently skip training classes because they see no future in which they can actually use their new skills.

VI. REPATRIATION AND MINE ERADICATION

Once the fighting stops, tens of thousands of Cambodians will begin leaving displaced persons camps inside the country and returning to their villages. Others, perhaps as many as 300,000, will return to Cambodia from the Thai border camps. They will leave the camps either spontaneously (mostly on foot) or in transport organized by the UNHCR and other international relief organizations. No one doubts that repatriation will bring with it a multitude of problems. Camp residents, especially those in the Khmer Rouge camps, fear that they will be stigmatized by those who chose to remain in Cambodia. Many will return to their villages to find their land farmed by others. Prime farming land will be in short supply. There will be a shortage of draft animals, farm implements, and fertilizers, as well as a lack of means to transport crops to market. People who have acquired professional skills in the camps are likely to encounter resentment from locally trained professionals as they compete for jobs. Many of the new arrivals will have no home, no village, and no immediate means of livelihood and may wander in search of work. Some will drift to towns or Phnom Penh, despite a likely ban on resettlement in urban areas. All of these factors means there will be mass movement across the country, and with the movement, the dangers posed by mines will increase.

In 1989, a Ford Foundation-funded survey of over 15,000 camp residents in Site 2, Site 8, and Site B found that the overwhelming majority wanted to take up farming on their return to Cambodia.[87] However, the survey also found that a significant number of camp residents (35 percent), primarily between the ages of 15 and 35, had either never farmed or had done so only during the Khmer Rouge period. John R. Rogge, a specialist in disaster

---

[87] J. Lynch, "Border Khmer: A Demographic Study of the Residents of Site 2, Site B, and Site 8," Ford Foundation, Bangkok, pp. 50-52.

research at the University of Manitoba, believes that this lack of farming experience among a sizeable number of camp residents will lead to further migration inside Cambodia. He states:

> They have no knowledge of seeds, of irrigation, of soils and weeds, or 'reading' the weather, the upcoming monsoon and its associated floods. They have little or no appreciation of the many risk-mitigating strategies that a farmer must regularly employ. Some of these needs can be taught, and are indeed being addressed by programming in the camps...Above all, rice farming requires a tenacity that is unlikely to be 'learned' by many youths who have lingered in idleness for several years in the camps. Moreover, most returnees are likely to have to occupy currently unused land; such land may need to be cleared and levelled, it may take several seasons to establish a good crop. Therefore, it is hypothesized that many who have indicated...a desire to return to Cambodia as farmers, may, once back, change their minds and seek other forms of employment. Unless income diversification programs are introduced in rural areas, many are likely to undertake a secondary migration to the towns.[88]

[88]See John R. Rogge, "Return to Cambodia," p. 123.

PROVINCE OF BIRTH AND INTENDED PROVINCE OF RETURN, SITE 2, SITE 8, SITE B POPULATIONS

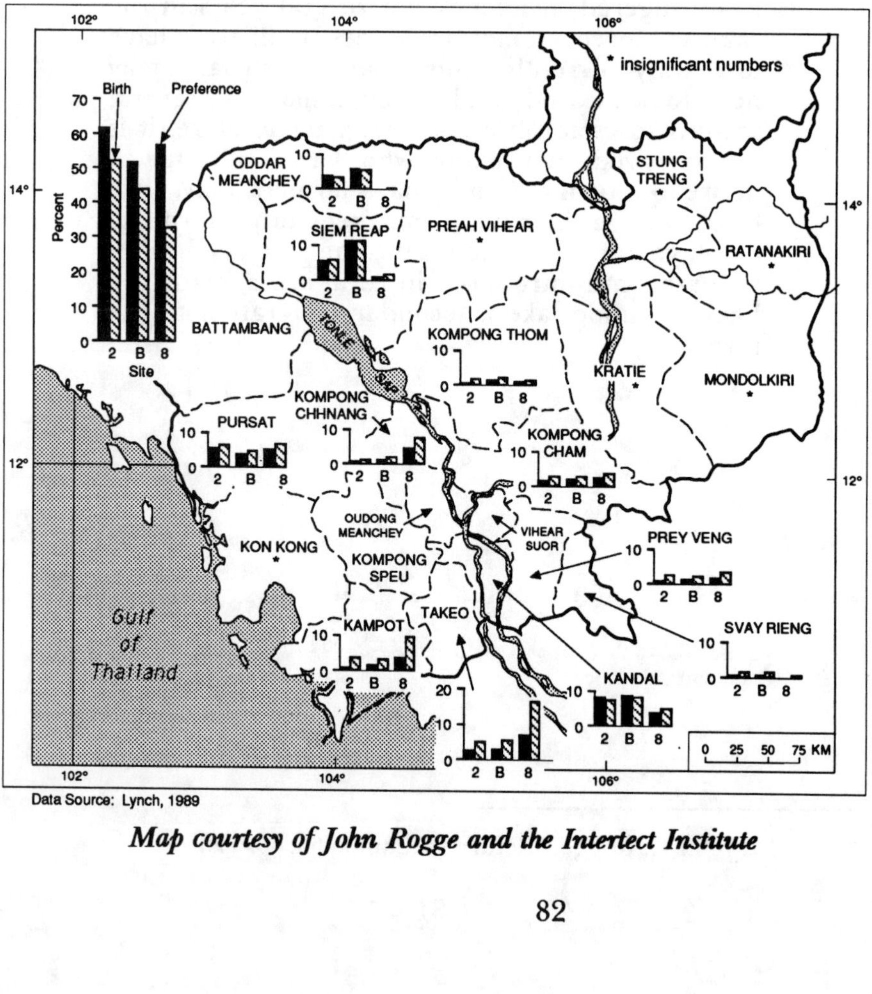

Data Source: Lynch, 1989

Map courtesy of John Rogge and the Intertect Institute

Most of the refugees are expected to settle in provinces closest to the Thai border where the concentration of mines is probably highest. According to the Ford Foundation survey, an average of 43 percent were born in Battambang province.[89] And a higher percentage, 57 percent, from the same camps indicated a preference for returning to Battambang. In contrast, no province other than Siem Reap was cited as a possible place of return by more than ten percent of any of the camp populations. The survey found that the high preference for Battambang province was due to "familiarity; people either originate from there, or they became familiar with it as a result of relocation during the Khmer Rouge era or during their transit to the border."[90] Battambang is also attractive because it is known as one the most fertile areas in Cambodia. In recent years, however, fewer and fewer fields in the province have been cultivated because of mines and increased fighting.[91]

Almost everyone familiar with the Cambodian conflict agrees that reclaiming settlement land in Cambodia will be a risky venture. In October 1990, United Nations Secretary General, Javier Pérez de Cuéllar, told the General Assembly:

> A serious problem which deserves the urgent attention of the international community is the presence of land mines and other anti-personnel devices, particularly in the western provinces of Cambodia to which most of the refugees and

---

[89]See J. Lynch, "Border Khmer: A Demographic Study of the Residents of Site 2, Site B, and Site 8."

[90]See J. R. Rogge, "Return to Cambodia," p. 112.

[91]This is also true of Banteay Meanchey province where only 115,000 of 215,000 hectares of land were cultivated because of mines, increased fighting, and a lack of draught animals, such as oxen. Interviews with Janetta Kwatia, UNICEF consultant on family food production programs, Phnom Penh, April 9, 1991 and the ICRC delegate, Battambang, April 11, 1991.

> displaced persons are expected to return. While mine-clearing activities can only take place following a peace accord, we have already begun a mine-awareness programme for the civilian population at the border, particularly women and children.[92]

More recently, a needs assessment team sent to Cambodia by the U.S. Agency for International Development, found that "Antipersonnel mines strewn everywhere in the country result in a high increase in injuries among agricultural civilians." The team concluded that "Rural areas must be de-mined immediately to prevent further injuries." A U.N. fact-finding mission sent to KPNLF and FUNCINPEC held areas in northwestern Cambodia in May 1991 to assess humanitarian assistance needs also concluded that land mines were a constant and serious threat to civilians:

> The mission was extremely aware of the security risks involved, particularly from land mines, and followed strict precautions to avoid any incidents. This involved proceeding in convoy only on main routes (usually in extremely poor condition) and moving freely and separately only within villages. Separate visits, especially to agricultural areas, were not possible. The mission witnessed first hand the grim and constant danger that mines pose to all populations in the areas visited.[93]

---

[92]"The Situation in Cambodia," Report of the Secretary-General of the United Nations (A/45/605, October 10, 1990, 45th Session, Agenda Item 32).

[93]"Report of U.N. Fact-finding Mission to KPNLF and FUNCINPEC Areas of Cambodia (18-22 May 1991)," Office of the Special Representative of the Secretary-General for Co-ordination of Cambodian Humanitarian Assistance Programmes, June 3, 1991, p. 2.

The U.N. team recommended that "Mine location/mapping programmes should commence with any relief programme, for the benefit of relief workers and local inhabitants. Safe routes should be mapped in advance for any relief workers."

These concerns were echoed, though with a greater note of futility, by the UNHCR mission chief in Phnom Penh, Bjorn Johansson, in an interview with our delegation in April 1991. "Mines will be cleared by people walking on them," he said, "but it will take a long time. The mines are an issue this country [Cambodia] must live with."

Such concern notwithstanding, there appears to be little consensus and considerable confusion within the United Nations as to how the problem can best be dealt with or, as Johansson suggests, whether it can be dealt with at all. The prevailing view among U.N. officials we interviewed in Bangkok and Phnom Penh was that land mine eradication was a military issue and thus one which could only be addressed once a cease-fire was agreed upon and respected by all sides. Some officials, while professing ignorance of the technicalities of mine eradication, nonetheless had strong opinions regarding the practicality, or even possibility, of instituting a mine eradication program in Cambodia. In contrast, doctors and relief workers who work daily with victims of mine blasts were of the opinion that something must be done immediately to remove the mines. ICRC delegates considered mines one of Cambodia's most serious public health problems. The regional director of Handicap International, Susan Walker, said: "We are only reacting to the situation by putting together the pieces after the fact. There is a real need for an initiative that strikes at the cause--that means removing the mines and destroying them before they destroy people."

The UNHCR Repatriation Plan

Although it is likely that with peace, many Cambodian refugees in Thailand will return spontaneously to Cambodia, the majority will most likely opt to return under the UNHCR repatriation plan. Given the probable scale of any return movement, the UNHCR has already begun working with UNBRO, ICRC, and relief organizations on the border in the preparation process.[94] Much of the logistical planning has focused on transportation, food rations, basic resettlement kits, and the location and security of reception centers inside Cambodia.

Once the UNHCR plan goes into operation, refugees will be moved to one or more preparation sites inside Thailand where registration, tracing services, medical examinations, and other preparatory services will be established. Before that, however, they will have chosen where they wish to be resettled in Cambodia. The land resettlement process, according to UNHCR mission chief Johansson, will be based on a "first come, first served" basis. The UNHCR will display large maps of Cambodia, based on 1:50,000 satellite photographs taken in early 1991, in the border camps and provide counseling on the agricultural potential of unoccupied land in the country. Refugees will then choose a plot of land and be issued a title.

From the preparation centers in Thailand refugees will be transported to reception centers inside Cambodia. The most likely mode will be by bus via the Aranyaprathet to Battambang road or the Prasat to Sam Rong road. About 10,000 refugees per week will be transported by bus, and possibly by rail, to six reception centers near the towns of Mongol Borei, Battambang, Sangke, Mong Russey, Pursat, and the capital, Phnom Penh. At that rate, and assuming no one returned spontaneously, it would take over seven months to repatriate all of the refugees.

Returnees will stay only a few days in the reception centers--long enough to be issued a food ration, water, and a resettlement kit that will include basic farming implements. They will then be "taken by truck to the farthest point on the road" and sent to their new homes, according to Johansson. Once the

---

[94]This "collaboration" has had its difficulties. Since the border camps are not under UNCHR jurisdiction, some relief workers have questioned how it can make plans for a population to which it largely has no access. In addition, refugees and relief workers alike have complained that the UNHCR has failed to consult refugees in the planning process.

returnees have located their plots of land, they can return to the reception centers for basic building supplies and food rations.

In May 1990, a task force comprised of representatives from the World Health Organization and six U.N. agencies travelled to western and southeastern Cambodia to assess the kinds of logistical support and assistance that would be needed for the repatriation of Cambodian refugees. Its report was unequivocal in its recognition of mines as a major obstacle to repatriation: "The mission has concluded that unless extensive physical preparations are carried out within a suitable time-frame *prior* to the conclusion of a political settlement, repatriation could entail catastrophic consequences due to the widespread presence of mines." With emphasis on key words, the report goes on to restate the magnitude of the situation relevant to repatriation: "The *de-mining* of access points from Thailand, routes to reception centres, designated settlement areas and agricultural land to be assigned to returnees is *crucial* to the *successful implementation* of the whole repatriation operation and should be dealt with in priority during the pre-operational phase."[95]

However, the report recommends an initial financial commitment of only $336,900 toward the mines problem, of which $296,000 would be for mine-awareness programs and the remaining $40,000 would be earmarked for a mine-detector dog pilot project.[96] It also calls for the development of a comprehensive de-mining plan based on information supplied by

---

[95]See "Report of Inter-Agency Mission to Thailand and Cambodia on Repatriation or Cambodian Refugees and Displaced Persons: 16 May - 10 June 1990 (Executive Summary)," United Nations, p. 2.

[96]A pilot project seems unnecessary as the Thai army, which is one of the major users of mine-detecting dogs, is well-qualified to report on their suitability for use in Cambodia. In addition, the U.N. mine clearance team in Afghanistan has worked in close cooperation with a project funded by the U.S. Agency for International Development that uses animals from Thailand and the United States. They, too, should be able to access if it is feasible to use dogs in Cambodia.

the Phnom Penh government and three resistance groups. It then recommends a three-phase operation to clear mines throughout Cambodia. The phases called for, in summary, are: (1) the demining and marking of access routes into Cambodia from Thailand and on routes to UNHCR reception centers; (2) the clearance of "designated settlement areas including agricultural land known to contain mines" to which the initial flow of refugees is sent; and (3) the clearance, "to the extent possible," of "surrounding areas known or suspected to contain mines."

So far, the U.N. report is the only official assessment of the land mine situation in Cambodia. However, it falls far short of addressing the real logistic and operational considerations that will be required to implement a large-scale mine eradication program in Cambodia.[97] To begin with, the budget for the mine-detecting dog project is unrealistic. As for the mines awareness programs, one of which is now operating in the Thai border camps, such efforts, though useful, have limited impact. Referring to the work of the U.N. mines awareness program, L.M.A.P., in the camps, one refugee in Site II jokingly said it was "like trying to teach the crocodile how to swim." Cambodian villagers and refugees know about the danger of mines, and they are aware that their rice paddies and fields may be mined, but they must work to survive and consequently have conditioned themselves to ignore the risks. Finally, the expectation that the four warring factions will turn over data on the location of minefields may turn out to be unrealistic. We found, for example, no indication that any faction possesses such records.

As for a policy leading to the actual eradication of land mines in Cambodia, it appears that such a program would be assigned to a U.N. peace-keeping force. Still, some U.N. officials,

---

[97]It should be noted that the inter-agency report did recognize that "The total actual extent and scope of the problem should...be confirmed by experts in the field of mine warfare." However, that the mission failed to enlist the services of such an expert is not encouraging given the gravity of the situation.

like Johansson, are not sanguine about the outcome of such an operation:

> According to the peace plan, each faction must turn over maps of mined areas and previous combatants will participate with UNTAC [the United Nations Transition Authority in Cambodia] troops in demining. I have serious doubts that this will work. This is a military task not a UNHCR task. Ideally it will take four to five months from a cease-fire for UNTAC to clear the mines, then the refugees can be let back in, but there isn't much trust in the process, and the U.N. is not getting the money it needs.[98]

Johansson and other UNHCR officials are in a "Catch 22" situation. They recognize the danger returnees face from mines, but they have neither the authority to stop the refugees from returning nor the mandate to clear the mines. Yet, once the war is over, they will be vested with the task of returning the refugees. The situation is further complicated by the fact that UNTAC, the agency assigned the task of locating and clearing the mines, does not even exist. "The point is you can never clear all the mines, but you can minimalize the problem" says Johansson. "You can first get a picture of where they were placed and then encourage people not to go to mined areas."

What will happen once the four warring factions reach a political settlement? One scenario is that once UNTAC is established, its military force will use flails (tank-like mine clearing machines) and other demining equipment to clear the major access routes to the reception centers and resettlement areas, thus paving the way for an organized repatriation four or five months later. Some observers fear, however, that once a settlement is

---

[98]Interview, Phnom Penh, April 9, 1991. In our view, the estimate of 4 to 5 month's time to eradicate land mines in Cambodia is unrealistic, particularly given the examples of clearance programs in Afghanistan and Vietnam.

reached, many refugees, up to 50,000 or more, will opt to return independently of any UNHCR-organized exercise.[99] There is also the question as to whether the resistance factions will agree to participate in the repatriation plan. The KPNLF, for example, have issued a statement criticizing the proposed reception centers as a means of "neutralizing the Cambodian resistance parties."[100]

Whatever the outcome, the problem of unexploded mines will remain. Gavin O'Keafe of CONCERN, a member of the team charged with setting up the reception centers, put it this way: "The sites themselves are undoubtedly safe. The problem begins when the people return to their homes and fields. I just can't see how anything can be done to survey or remove the mines in the present situation. There are four armies, bandits, and constantly shifting forces and, therefore, mines. What can be done?"[101]

A Mine Eradication Program

The nature of mine warfare has changed considerably with the advent of randomly disseminated mines, particularly anti-personnel mines. Durable, lightweight, and cheap, they can be easily carried and deployed rapidly and in large numbers, not only to protect military positions but to discourage enemy soldiers and civilians from entering certain areas. Those who suffer most from the use of these mines are farming and herding communities.

---

[99]See J.R. Rogge, "Return to Cambodia," p. 3. According to Rogge, one of the reasons refugees may choose to return spontaneously is because they seek to avoid "protracted processing times and/or movement through a series of 'transit' or 'reception' centres."

[100]See KPNLF memorandum on the question of repatriation, October 25, 1990, p. 3.

[101]Interview, Battambang, April 11, 1991.

Until recently, little international attention has been given to developing land mine eradication techniques for large expanses of pasture or agricultural land. This is due, in part, to the fact that mine clearing technologies have traditionally been geared toward military purposes. When military units clear land mines it is normally to allow the passage of soldiers and their equipment over an area of ground. This is commonly accomplished by a breaching operation, or the clearance of a tract of land no greater than is necessary to achieve the immediate military objectives.

General recognition within the international aid community of the land mine issue as one relevant to their role has been slow. Even now, there is a reluctance in some circles to accept either the humanitarian nature of the problem or its scale and seriousness. An example of the latter tendency to minimize the impact of land mines on communities is found in the September 1989 UNHCR Report for Repatriation Planning for Afghan Refugees which stated that the problem of mines has been exaggerated. Subsequent, as well as preceding, accounts have shown this statement to be, at best, irresponsible misinformation.

Nonetheless, it was in Afghanistan that the concept of humanitarian-based land mine eradication was first established. The conceptual underpinnings of this program were based on three fundamental precepts.[102] First, the program should be directed at stopping the loss of life and injury caused by land mines among non-combatants and livestock, as well as returning mined land to its peacetime use. Second, the program's aims should be strictly humanitarian and not military or political. Therefore, no mines should be handed over to military units. To ensure absolute neutrality, the program should be administered by the United Nations. Finally, all mines detected as a result of the program must be destroyed either *in situ* or, where this is

[102]Donors and controlling agencies must ensure that these precepts are a contractual requirement and that infringement will result in the termination of funding and support.

impractical, they should be removed, taken to a safe location, and destroyed.

Based on the existing program in Afghanistan, we recommend a four-phase mine eradication program for Cambodia. It is important to recognize that it will be expensive and will require some military assistance.

Phase One (community survey). Similar to the 13-province survey conducted in Afghanistan by the Mines Advisory Group in 1989-1990[103], the Cambodian survey would focus on collating first-hand data at the community level, interviewing mine casualties and combatants, and monitoring mine-affected areas. (See Appendix C) The aim would be to provide an overall picture of the problem detailed enough to be used as a basis for planning, budget, and equipment estimates and as a guide for agencies involved in repatriation, prostheses, and related programs. Ideally, this survey should begin immediately. Some aspects of the survey may require the cooperation of combatant groups. However, a failure to gain their cooperation should not hinder the commencement of the survey.

Phase Two (technical survey). Specific areas for mine eradication operations must be defined, mapped, and marked by, or under the supervision of, specialists. This phase could begin in uncontested areas without undue delay. Technical surveying of a larger scale would depend on an end to hostilities and formal operational agreements between all parties involved.

Phase Three (eradication in priority areas). Immediately following the technical survey, land mines would be destroyed, under the auspices of the United Nations, in selected priority areas.

Phase Four (long-term eradication). The major tasks during this phase would be eradicating mines from rice paddies,

---

[103]See Mines Advisory Group, *Report of the Afghanistan Mines Survey*, February 1991.

pastures, fields, villages, pathways, shorelines, and previously uncleared border areas. This phase would require a reasonable level of political stability and cooperation.

Aims and objectives need to be clearly defined as soon as possible, and potential donors need to be involved in the planning process to ensure that they are aware of the level of commitment and support required. Above all, the program should be planned, implemented, and directed by experts in mine eradication.

Community land mine surveys are not an academic process; their prime objective is to establish yardsticks for planning purposes. In a field of operations where equipment is expensive and not always readily available, it is essential to order mine eradication equipment in advance and this will cut capital outlay considerably. Experienced field staff for mine eradication command high prices; salaries will be high and the task of locating suitably qualified consultants will take time. One value of a thorough survey is that it helps to identify the scale and type of operation needed and to establish training, staffing, and funding requirements in advance.

Technical Considerations

While a mine survey in Cambodia will give a more complete picture of the technical factors to be considered, there are sufficient data to begin research and preparation to deal with some of the problems unique to the Cambodian situation.

■ **Operations in paddy-fields**: It may be that operations in these fields must be confined to the dry season, but this would considerably lengthen the process. Research needs to be undertaken to determine (1) the effects on mines of prolonged immersion on mines and other devices in rice paddies and the safest and most efficient means of eradication; (2) the realistic potential for year-round operations in wetland areas; and (3) the possibility and realistic potential of localized drainage as a precursor to eradication operations.

■ **Eradication in jungle**: Many mines in Cambodia have been disseminated in jungle areas where vegetation growth is rapid and dense. Some of the devices may have been laid up to 12 years ago in areas now heavily overgrown.

■ **Making villages safe**: Until further information is available on specific methods of land mine deployment in villages, the types of devices employed, and the frequency with which booby-traps and improvised explosive devices have been used, no task assessment can be made. It is obvious that the eradication of mines from villages and their immediate surroundings must be a priority. Although there does not appear to be evidence of widespread use of booby-traps in houses, the use of mines near sources of fresh water and on paths within villages is a major concern and will require extreme caution on the part of search and clearance teams.

Funding

The cost for removing land mines in Cambodia will be proportional to the investment made in supplying mines to the combat factions. To carry the eradication program through to completion will require a large investment--many millions of dollars. Clearly, Cambodia has neither the economic nor organizational infrastructure to support such a large-scale undertaking in the immediate or foreseeable future. It must therefore fall to the international community to provide the funding and organization, at least in the short term.

Since the outbreak of the Cambodian civil war nearly 12 years ago, several countries have provided military support to the four combatant factions, the Phnom Penh Government, Khmer Rouge, Khmer People's National Liberation Front or Armee Nationale Sihanoukienne. All of these factions and their supporters have claimed to have had the future of Cambodia and the rights of the Khmers at heart. It should not require, therefore, any change in policy by those countries to fund a program so clearly beneficial to all Khmers, regardless of their politics. Such a commitment by China, the United States, the Soviet Union, Vietnam, the United Kingdom, and the ASEAN nations, all of which have been heavily involved in arming and training the combatant factions, would be likely to ensure sufficient funds to institute a realistic large-scale eradication program.

The Cost of Failure

Given the already devastating effects of land mines in Cambodia, there are some obvious, if unpalatable, questions that must be asked: What if there is no lasting cease-fire, no durable political settlement? What if the mines continue to be laid for another two decades?

The answer is simple but appalling. Assuming a continuing supply of mines to all four factions at the present rate, and assuming that present strategies are unlikely to change, more and more tracts of land in Cambodia will become uninhabitable, as areas along the Thai border already are. Continued targeting of villages, rice paddies, forest margins, river banks, key logistic routes and railways, combined with natural flora growth, seasonal flooding, and a total lack of record-keeping, may make any repatriation program unthinkable.

VII. OBSERVATIONS AND RECOMMENDATIONS

Indiscriminate Use of Land Mines

All sides in the Cambodian conflict have used land mines without regard for their effects on the civilian population. None of the warring parties have routinely kept records or maps of the location of mines, nor have they warned civilians of mines laid in areas frequented by noncombatants. In some cases, the resistance forces, especially the Khmer Rouge, have knowingly placed land mines in civilian areas to terrorize the local population or to control their movement. Moreover, both government and resistance forces often fail to remove mines once their military objective has been fulfilled. This has particularly been the case along pathways frequented by civilians.

Statistics on war casualties from hospitals in Cambodia and the Thai border camps show that, during 1990 and the first three months of 1991, land mines injured more combatants and noncombatants than any other weapon. Of the injured, over half were civilians, including a significant number of women and children. In addition, Khmer and foreign surgeons report that this has been the trend for several years. These statistics, coupled with accounts from government soldiers and resistance fighters, clearly indicate that all sides have flagrantly disregarded the rule of proportionality, which holds that civilian casualties and damage to civilian objects should not be out of proportion to the military advantages anticipated.

Since the start of the Cambodian conflict, several nations, including those who claim an interest in the cessation of hostilities, have supplied huge stocks of land mines and other devices to either the Phnom Penh government or one or all, of the three resistance factions--the Khmer Rouge, KPNLF, and ANS. Those most involved have been China, the Soviet Union, Singapore, Thailand, Vietnam, and the United States. In addition, at least two nations, China and the United Kingdom, are, or have been, involved in training Cambodian resistance factions in the use of mines and explosives against civilian as well as military targets. All of these nations have a moral obligation and responsibility to intervene when their clients use mines in a manner that endangers the lives and safety of civilians.

- Given the present cease-fire, all parties to the conflict should desist from laying land mines and other devices, and their foreign sponsors should stop supplying them and call publicly for an end to their use in Cambodia. This international embargo should include all mines, including those in warehouses not yet issued to combat units.

- As the Phnom Penh government and the resistance factions negotiate a political settlement to the conflict, they should make the clearance and disposal of land mines a high priority. If any records of minefields exist, all parties to the conflict should make them available to international and private agencies involved in humanitarian mine eradication.

- The manufacturers of land mines now in use in Cambodia should provide technical information to eradication teams to aid them in their work.

- Once a political settlement is reached, the Supreme National Council, or whatever transitional government is installed, should make it a punishable crime for anyone other than authorized personnel to deactivate and store mines.

Mine Eradication

Land mines and other devices that detonate on contact are indiscriminate weapons that can maim or kill for years, even decades, after they are laid. In Cambodia, as in most developing countries, mines pose a great danger to farming communities, cause environmental destruction, and hinder economic growth. The widespread presence and density of land mines in Cambodia must be considered a humanitarian emergency, separate from, and regardless of, the other crises facing the Khmer people.

- The international community, under the auspices of the United Nations, must be prepared to underwrite a mine eradication program and preparations should begin immediately. Nations that have supplied mines to Cambodia's four warring parties have a moral obligation and responsibility to share in the costs of land mine eradication. The U.S. government should take the lead in providing funds for mine surveying and eradication in Cambodia and encourage other nations to contribute.

- The program should be carried out under the auspices of the United Nations, and its objectives should be humanitarian rather than military. Measures must be taken to prevent combat units or other groups from removing mines and stockpiling them for possible later use. At the same time, all existing stockpiles should be turned over, along with other munitions, to the United Nations team and destroyed.

Repatriation

There are now 300,000 Cambodians in refugee camps in Thailand, with an additional 180,000 in displaced persons camps inside Cambodia. These people will eventually leave their encampments and return to their villages either spontaneously or under a repatriation program organized by the United Nations High Commission for Refugees. Most will return to the war-torn provinces in the northwest and southwest where land mines are most prevalent.

In recent years, the resistance factions in the Thai border camps have begun establishing satellite camps inside Cambodia. Resistance leaders have encouraged or forced Khmers coming to the border from inside Cambodia and people from the Thai border camps to settle in these areas. They have built hospitals and roads. In 1990, several leaders of the different factions declared that they had drafted their own repatriation plans, and that they saw little need for an operation carried out under the auspices of the United Nations.

The ICRC and other relief organizations on the border have publicly denounced any repatriation of the border population until a comprehensive political settlement and suspension of hostilities has been reached. They have also said that camp residents should be given freedom to determine how and where in Cambodia they wish to go. The ICRC has repeatedly warned of the widespread presence of mines inside Cambodia and the dangers they will pose to returnees.

- Under no circumstances should the border population or the displaced inside Cambodia be encouraged or pushed to move into areas where they are endangered by mines. A mines survey in potential settlement areas inside Cambodia should begin immediately (see Chapter VI and Appendix C).

- The international community should encourage the Phnom Penh government and the resistance factions to cooperate with surveying teams. The mines survey should also be a point of discussion when representatives of the Supreme National Council and the United Nations High Commissioner for Refugees meet later this year.

Medical Treatment and Rehabilitation

Cambodia's health care delivery system is unable to cope with the demands of a population plagued by disease and war. There is a shortage of blood, medicines, and properly trained medical personnel in hospitals and clinics throughout the country. Given these conditions, victims of mine blasts are particularly vulnerable. They suffer from injuries that often require large amounts of blood, antibiotics, and extended stays in hospital. After discharge from hospital, many victims will require physical therapy and a prosthetic device to enable them to live normal lives. Most mine victims, however, will never receive these services.

Khmer and foreign doctors and health workers are struggling to provide medical care to mine victims and other war victims under difficult circumstances, and they deserve the support of their colleagues worldwide. Several medical and relief organizations merit recognition, including the Cambodian Red Cross, the International Committee of the Red Cross, Handicap International, American Friends Service Committee, Red Cross societies, Médecins sans Frontieres, and Médecins du Monde. Still, more needs to be done to aid victims of this man-made disaster.

- Since 1979, Cambodia's small medical community has been ostracized by a large segment of the international medical community for political reasons. This isolation should end immediately. To aid mine victims, international and national medical organizations and medical schools could establish supplemental training and exchange programs in surgery, including the use of external fixtures and physical therapy, for Khmer doctors, nurses, and health workers.

Land Mines Protocol

The most ambitious attempt to control the use of mines and the harm they inflict on civilians is the Land Mines Protocol (see Chapter I and Appendix A). Another protocol contained in the same convention, Protocol I, also prohibits the use of weapons, including mines, whose primary objective "is to injure by fragments which in the human body escape detection by X-ray."

It has been nearly 10 years since these protocols were adopted, and there is little to show for it. To begin with, the protocols only apply to international armed conflicts. Yet, most of the recent conflicts in which combatants have used (or continue to use) land mines with little or no regrard for civlian lives have been internal. These include Afghanistan, Somalia, Angola, Uganda, El Salvador, Guatemala, Ethiopia, Sudan, Nicaragua, Mozambique, and Burma. Moreover, most combatants are unaware of the protocols. Our delegation found, for instance, that the only people we interviewed in Cambodia and Thailand (including U.N. and relief officials, soldiers, and resistance fighters) who were aware of the protocols were ICRC delegates.

Another major weakness of the Land Mines Protocol is the fact that it gives too much discretion to military commanders and soldiers who must make decisions on the deployment of mines under battle conditions (see Chapter IV). Soldiers, for instance, who are retreating quickly from behind enemy lines may decide to mine escape routes, even though they are well aware that civilians use them every day. Moreover, it is difficult at times to discern whether the Land Mines Protocol is directed at protecting civilians or supplying advance vindication for unjustifiable military actions. For example, the Protocol states that "Effective advance warning shall be given of any delivery or dropping of remotely delivered mines which may affect the civilian population, *unless circumstances do not permit*. Soldiers are only required to take *feasible* precautions to protect civilians [emphases added]. Those qualifiers, as well as several others that appear in the Protocol, enable military field commanders--and ultimately governments--to evade legal responsibility if civilians die from the mines they have laid.

Protocol I also fails to address reality. It bans the use of plastic shrapnel but today, many mines are being manufactured with hard plastic casing (see Chapter III). Even though manufacturers can argue that the plastic casing is there to make the mine lighter and more durable, it also can turn into shrapnel after explosion and become embedded in wounds. As noted in Chapter IV, shards of plastic casing that enter wounds are frequently undetectable by X-ray.

- Asia Watch and Physicians for Human Rights urge the United Nations and the ICRC to re-evaluate the effectiveness of Protocols I and II of the 1981 Convention. In doing so, these bodies should seek advice from representatives of relief, medical, de-mining, and military organizations. They should base their review on epidemiologic data on the use of land mines and their effects on civilian populations in countries that have recently experienced or are in the grip of international or internal conflict. If this data does not exist, they should obtain it from hospitals and Red Cross missions.

- Among the options taken under review, the United Nations and the International Committee of the Red Cross should consider an unconditional ban on the manufacture, possession, transfer, sale and use of land mines and other devices that detonate on contact in all international and internal conflicts. While such a law might not entirely eliminate the use of mines, it would stigmatize them in much the same way that chemical weapons are now vilified.

APPENDIX A

Convention on Prohibitions or Restrictions on the Use of Certain Conventional Weapons Which May be Deemed to be Excessively Injurious or to Have Indiscriminate Effects

The High Contracting Parties,

Recalling that every State has the duty, in conformity with the Charter of the United Nations, to refrain in its international relations from the threat or use of force against the sovereignty, territorial integrity or political independence of any State, or in any other manner inconsistent with the purposes of the United Nations,

Further recalling the general principle of the protection of the civilian population against the effects of hostilities,

Basing themselves on the principle of international law that the right of the parties to an armed conflict to choose methods or means of warfare is not unlimited, and on the principle that prohibits the employment in armed conflicts of weapons, projectiles and material and methods of warfare of a nature to cause superfluous injury or unnecessary suffering,

Also recalling that it is prohibited to employ methods or means of warfare which are intended, or may be expected, to cause widespread, long-term and severe damage to the natural environment,

Confirming their determination that in cases not covered by this Convention and its annexed Protocols or by other international agreements, the civilian population and the combatants shall at all times remain under the protection and authority of the principles of international law derived from established custom, from the principles of humanity and from the dictates of public conscience,

Desiring to contribute to international détente, the ending of the arms race and the building of confidence among States, and hence to the realization of the aspiration of all peoples to live in peace,

Recognizing the importance of pursuing every effort which may contribute to progress towards general and complete disarmament under strict and effective international control,

Reaffirming the need to continue the codification and progressive development of the rules of international law applicable in armed conflict,

Wishing to prohibit or restrict further the use of certain conventional weapons and believing that the positive results achieved in this area may facilitate the main talks on disarmament with a view to putting an end to the production, stockpiling and proliferation of such weapons,

Emphasizing the desirability that all States become parties to this Convention and its annexed Protocols, especially the military significant States,

Bearing in mind that the General Assembly of the United Nations and the United Nations Disarmament Commission may decide to examine the question of a possible broadening of the scope of the prohibitions and restrictions contained in this Convention and its annexed Protocols,

Further bearing in mind that the Committee on Disarmament may decide to consider the question of adopting further measures to prohibit or restrict the use of certain conventional weapons,

Have agreed as follows:

Article 1 – *Scope of application*

This Convention and its annexed Protocols shall apply in the situations referred to in Article 2 common to the Geneva Coventions of 12 August 1949 for the Protection of War Victims, including any situation described in paragraph 4 of Article 1 of Additional Protocol I to these Conventions.

Article 2 – *Relations with other international agreements*

Nothing in this Convention or its annexed Protocols shall be interpreted as detracting from other obligations imposed upon the High Contracting Parties by international humanitarian law applicable in armed conflict.

Article 3 – *Signature*

This Convention shall be open for signature by all States at United Nations Headquarters in New York for a period of twelve months from 10 April 1981.

Article 4 – *Ratification, acceptance, approval or accession*

1. This Convention is subject to ratification, acceptance or approval by the Signatories. Any State which has not signed this Convention may accede to it.

2. The instruments of ratification, acceptance, approval or accession shall be deposited with the Depositary.

3. Expressions of consent to be bound by any of the Protocols annexed to this Convention shall be optional for each State, provided that at the time of the deposit of its instrument of ratification, acceptance or approval of this Convention or of accession thereto, that State shall notify the Depositary of its consent to be bound by any two or more of these Protocols.

4. At any time after the deposit of its instrument of ratification, acceptance or approval of this Convention or of accession thereto,

a State may notify the Depositary of its consent to be bound by any annexed Protocol by which it is not already bound.

5. Any Protocol by which a High Contracting Party is bound shall for that Party form an integral part of this Convention.

Article 5 – Entry into force

1. This Convention shall enter into force six months after the date of deposit of the twentieth instrument of ratification, acceptance, approval or accession.

2. For any State which deposits its instrument of ratification, acceptance, approval or accession after the date of the deposit of the twentieth instrument of ratification, acceptance, approval or accession, this Convention shall enter into force six months after the date on which that State has deposited its instrument of ratification, acceptance, approval or accession.

3. Each of the Protocols annexed to this Convention shall enter into force six months after the date by which twenty States have notified their consent to be bound by it in accordance with paragraph 3 or 4 of Article 4 of this Convention.

4. For any State which notifies its consent to be bound by a Protocol annexed to this Convention after the date by which twenty States have notified their consent to be bound by it, the Protocol shall enter into force six months after the date on which that State has notified its consent so to be bound.

Article 6 – Dissemination

The High Contracting Parties undertake, in time of peace as in time of armed conflict, to disseminate this Convention and those of its annexed Protocols by which they are bound as widely as possible in their respective countries and, in particular, to include the study thereof in their programmes of military instruction, so that those instruments may become known to their armed forces.

Article 7 – Treaty relations upon entry into force of this Convention

1. When one of the parties to a conflict is not bound by an annexed Protocol, the parties bound by this Convention and that annexed Protocol shall remain bound by them in their mutual relations.

2. Any High Contracting Party shall be bound by this Convention and any Protocol annexed thereto which is in force for it, in any situation contemplated by Article 1, in relation to any State which is not a party to this Convention or bound by the relevant annexed Protocol, if the latter accepts and applies this Convention or the relevant Protocol, and so notifies the Depositary.

3. The Depositary shall immediately inform the High Contracting Parties concerned of any notification received under paragraph 2 of this Article.

4. This Convention, and the annexed Protocols by which a High Contracting Party is bound, shall apply with respect to an armed conflict against that High Contracting Party of the type referred to in Article 1, paragraph 4, of Additional Protocol I to the Geneva Conventions of 12 August 1949 for the Protection of War Victims:

(*a*) where the High Contracting Party is also a party to Additional Protocol I and an authority referred to in Article 96, paragraph 3, of that Protocol has undertaken to apply the Geneva Conventions and Additional Protocol I in accordance with Article 96, paragraph 3, of the said Protocol, and undertakes to apply this Convention and the relevant annexed Protocols in relation to that conflict; or

(*b*) where the High Contracting Party is not a party to Additional Protocol I and an authority of the type referred to in subparagraph (*a*) above accepts and applies the obligations of the Geneva Conventions and of this Convention and the relevant annexed Protocols in relation to that conflict. Such an acceptance and application shall have in relation to that conflict the following effects:

(i) the Geneva Conventions and this Convention and its relevant annexed Protocols are brought into force for the parties to the conflict with immediate effect;

(ii) the said authority assumes the same rights and obligations as those which have been assumed by a High Contracting Party to the Geneva Conventions, this Convention and its relevant annexed Protocols; and

(iii) the Geneva Conventions, this Convention and its relevant annexed Protocols are equally binding upon all parties to the conflict.

The High Contracting Party and the authority may also agree to accept and apply the obligations of Additional Protocol I to the Geneva Conventions on a reciprocal basis.

Article 8 — *Review and amendments*

1. (*a*) At any time after the entry into force of this Convention any High Contracting Party may propose amendments to this Convention or any annexed Protocol by which it is bound. Any proposal for an amendment shall be communicated to the Depositary, who shall notify it to all the High

Contracting Parties and shall seek their views on whether a conference should be convened to consider the proposal. If a majority, that shall not be less than eighteen of the High Contracting Parties so agree, he shall promptly convene a conference to which all High Contracting Parties shall be invited. States not parties to this Convention shall be invited to the conference as observers.

(*b*) Such a conference may agree upon amendments which shall be adopted and shall enter into force in the same manner as this Convention and the annexed Protocols, provided that amendments to this Convention may be adopted only by the High Contracting Parties and that amendments to a specific annexed Protocol may be adopted only by the High Contracting Parties which are bound by that Protocol.

2. (*a*) At any time after the entry into force of this Convention any High Contracting Party may propose additional protocols relating to other categories of conventional weapons not covered by the existing annexed Protocols. Any such proposal for an additional protocol shall be communicated to the Depositary, who shall notify it to all the High Contracting Parties in accordance with subparagraph 1 (*a*) of this Article. If a majority, that shall not be less than eighteen of the High Contracting Parties so agree, the Depositary shall promptly convene a conference to which all States shall be invited.

(*b*) Such a conference may agree, with the full participation of all States represented at the conference, upon additional protocols which shall be adopted in the same manner as this Convention, shall be annexed thereto and shall enter into force as provided in paragraphs 3 and 4 of Article 5 of this Convention.

3. (*a*) If, after a period of ten years following the entry into force of this Convention, no conference has been convened in accordance with subparagraph 1 (*a*) or 2 (*a*) of this Article, any High Contracting Party may request the Depositary to convene a conference to which all High Contracting Parties shall be invited to review the scope and operation of this Convention and the Protocols annexed thereto and to consider any proposal for amendments of this Convention or of the existing Protocols. States not parties to this Convention shall be invited as observers to the conference. The conference may agree upon amend-

ments which shall be adopted and enter into force in accordance with subparagraph 1 (*b*) above.

(*b*) At such conference consideration may also be given to any proposal for additional protocols relating to other categories of conventional weapons not covered by the existing annexed Protocols. All States represented at the conference may participate fully in such consideration. Any additional protocols shall be adopted in the same manner as this Convention, shall be annexed thereto and shall enter into force as provided in paragraphs 3 and 4 of Article 5 of this Convention.

(*c*) Such a conference may consider whether provision should be made for the convening of a further conference at the request of any High Contracting Party if, after a similar period to that referred to in subparagraph 3 (*a*) of this Article, no conference has been convened in accordance with subparagraph 1 (*a*) or 2 (*a*) of this Article.

Article 9 – *Denunciation*

1. Any High Contracting Party may denounce this Convention or any of its annexed Protocols by so notifying the Depositary.

2. Any such denunciation shall only take effect one year after receipt by the Depositary of the notification of denunciation. If, however, on the expiry of that year the denouncing High Contracting Party is engaged in one of the situations referred to in Article 1, the Party shall continue to be bound by the obligations of this Convention and of the relevant annexed Protocols until the end of the armed conflict or occupation and, in any case, until the termination of operations connected with the final release, repatriation or re-establishment of the persons protected by the rules of international law applicable in armed conflict, and in the case of any annexed Protocol containing provisions concerning situations in which peace-keeping, observation or similar functions are performed by United Nations forces or missions in the area concerned, until the termination of those functions.

3. Any denunciation of this Convention shall be considered as also applying to all annexed Protocols by which the denouncing High Contracting Party is bound.

4. Any denunciation shall have effect only in respect of the denouncing High Contracting Party.

5. Any denunciation shall not affect the obligations already incurred, by reason of an armed conflict, under this Convention and its annexed Protocols by such denouncing High Contracting

Party in respect of any act committed before this denunciation becomes effective.

Article 10 – *Depositary*

1. The Secretary-General of the United Nations shall be the Depositary of this Convention and of its annexed Protocols.

2. In addition to his usual functions, the Depositary shall inform all States of:

(*a*) signatures affixed to this Convention under Article 3;

(*b*) deposits of instruments of ratification, acceptance or approval of or accession to this Convention deposited under Article 4;

(*c*) notifications of consent to be bound by annexed Protocols under Article 4;

(*d*) the dates of entry into force of this Convention and of each of its annexed Protocols under Article 5; and

(*e*) notifications of denunciation received under Article 9, and their effective date.

Article 11 – *Authentic texts*

The original of this Convention with the annexed Protocols, of which the Arabic, Chinese, English, French, Russian and Spanish texts are equally authentic, shall be deposited with the Depositary, who shall transmit certified true copies thereof to all States.

* * *

Protocol on Non-Detectable Fragments (Protocol I)

It is prohibited to use any weapon the primary effect of which is to injure by fragments which in the human body escape detection by X-rays.

* * *

Protocol on Prohibitions or Restrictions on the Use of Mines, Booby Traps and Other Devices (Protocol II)

Article 1 – *Material scope of application*

This Protocol relates to the use on land of the mines, booby-traps and other devices defined herein, including mines laid to interdict beaches, waterway crossings or river crossings, but does not apply to the use of anti-ship mines at sea or in inland waterways.

Article 2 – Definitions

For the purpose of this Protocol:

1. 'Mine' means any munition placed under, on or near the ground or other surface area and designed to be detonated or exploded by the presence, proximity or contact of a person or vehicle, and 'remotely delivered mine' means any mine so defined delivered by artillery, rocket, mortar or similar means or dropped from an aircraft.

2. 'Booby-trap' means any device or material which is designed, constructed or adapted to kill or injure and which functions unexpectedly when a person disturbs or approaches an apparently harmless object or performs an apparently safe act.

3. 'Other devices' means manually-emplaced munitions and devices designed to kill, injure or damage and which are actuated by remote control or automatically after a lapse of time.

4. 'Military objective' means, so far as objects are concerned, any object which by its nature, location, purpose or use makes an effective contribution to military action and whose total or partial destruction, capture or neutralization, in the circumstances ruling at the time, offers a definite military advantage.

5. 'Civilian objects' are all objects which are not military objectives as defined in paragraph 4.

6. 'Recording' means a physical, administrative and technical operation designed to obtain, for the purpose of registration in the official records, all available information facilitating the location of minefields, mines and booby-traps.

Article 3 – General restrictions of the use of mines, booby-traps and other devices

1. This Article applies to:
   (*a*) mines;
   (*b*) booby-traps; and
   (*c*) other devices.

2. It is prohibited in all circumstances to direct weapons to which this Article applies, either in offence, defence or by way of reprisals, against the civilian population as such or against individual civilians.

3. The indiscriminate use of weapons to which this Article applies is prohibited. Indiscriminate use is any placement of such weapons:
   (*a*) which is not on, or directed at, a military objective; or

(*b*) which employs a method or means of delivery which cannot be directed at a specific military objective; or

(*c*) which may be expected to cause incidental loss of civilian life, injury to civilians, damage to civilian objects, or a combination thereof, which would be excessive in relation to the concrete and direct military advantage anticipated.

4. All feasible precautions shall be taken to protect civilians from the effects of weapons to which this Article applies. Feasible precautions are those precautions which are practicable or practically possible taking into account all circumstances ruling at the time, including humanitarian and military considerations.

Article 4 – Restrictions on the use of mines other than remotely delivered mines, booby-traps and other devices in populated areas

1. This Article applies to:

(*a*) mines other than remotely delivered mines;

(*b*) booby-traps; and

(*c*) other devices.

2. It is prohibited to use weapons to which this Article applies in any city, town, village or other area containing a similar concentration of civilians in which combat between ground forces is not taking place or does not appear to be imminent, unless either:

(*a*) they are placed on or in the close vicinity of a military objective belonging to or under the control of an adverse party; or

(*b*) measures are taken to protect civilians from their effects, for example, the posting of warning signs, the posting of sentries, the issue of warnings or the provision of fences.

Article 5 – Restrictions on the use of remotely delivered mines

1. The use of remotely delivered mines is prohibited unless such mines are only used within an area which is itself a military objective or which contains military objectives, and unless:

(*a*) their location can be accurately recorded in accordance with Article 7(1)(*a*); or

(*b*) an effective neutralizing mechanism is used on each such mine, that is to say, a self-actuating mechanism which is designed to render a mine harmless or cause it to destroy itself when it is anticipated that the mine will no longer serve the military purpose for which it was placed in position, or a remotely-controlled mechanism which is designed to render harmless or destroy a mine when the mine no longer serves the military purpose for which it was placed in position.

2. Effective advance warning shall be given of any delivery or dropping of remotely delivered mines which may affect the civilian population, unless circumstances do not permit.

Article 6 – *Prohibition on the use of certain booby-traps*

1. Without prejudice to the rules of international law applicable in armed conflict relating to treachery and perfidy, it is prohibited in all circumstances to use:

(*a*) any booby-trap in the form of an apparently harmless portable object which is specifically designed and con-constructed to contain explosive material and to detonate when it is disturbed or approached, or

(*b*) booby-traps which are in any way attached to or associated with:

(i) internationally recognized protective emblems, signs or signals;

(ii) sick, wounded or dead persons;

(iii) burial or cremation sites or graves;

(iv) medical facilities, medical equipment, medical supplies or medical transportation;

(v) children's toys or other portable objects or products specially designed for the feeding, health, hygiene, clothing or education of children;

(vi) food or drink;

(vii) kitchen utensils or appliances except in military establishments, military locations or military supply depots;

(viii) objects clearly of a religious nature;

(ix) historic monuments, works of art or places or worship which constitute the cultural or spiritual heritage of peoples;

(x) animals or their carcasses.

2. It is prohibited in all circumstances to use any booby-trap which is designed to cause superfluous injury or unnecessary suffering.

Article 7 – *Recording and publication of the location of minefields, mines and booby-traps*

1. The parties to a conflict shall record the location of:

(*a*) all pre-planned minefields laid by them; and

(*b*) all areas in which they have made large-scale and pre-planned use of booby-traps.

2. The parties shall endeavour to ensure the recording of the location of all other minefields, mines and booby-traps which they have laid or placed in position.

3. All such records shall be retained by the parties who shall:

(*a*) immediately after the cessation of active hostilities:

(i) take all necessary and appropriate measures, including the use of such records, to protect civilians from the effects of minefields, mines and booby-traps; and either

(ii) in cases where the forces of neither party are in the territory of the adverse party, make available to each other and to the Secretary-General of the United Nations all information in their possession concerning the location of minefields, mines and booby-traps in the territory of the adverse party; or

(iii) once complete withdrawl of the forces of the parties from the territory of the adverse party has taken place, make available to the adverse party and to the Secretary-General of the United Nations all information in their possession concerning the location of minefields, mines and booby-traps in the territory of the adverse party;

(*b*) when a United Nations force or mission performs functions in any area, make available to the authority mentioned in Article 8 such information as is required by that Article;

(*c*) whenever possible, by mutual agreement, provide for the release of information concerning the location of minefields, mines and booby-traps, particularly in agreements governing the cessation of hostilities.

Article 8 – *Protection of United Nations forces and missions from the effects of minefields, mines and booby-traps*

1. When a United Nations force or mission performs functions of peacekeeping, observation or similar functions in any area, each party to the conflict shall, if requested by the head of the United Nations force or mission in that area, as far as it is able:

(*a*) remove or render harmless all mines or booby-traps in that area;

(*b*) take such measures as may be necessary to protect the force or mission from the effects of minefields, mines and booby-traps while carrying out its duties; and

(*c*) make available to the head of the United Nations force or mission in that area, all information in the party's possession concerning the location of minefields, mines and booby-traps in that area.

2. When a United Nations fact-finding mission performs functions in any area, any party to the conflict concerned shall provide protection to that mission except where, because of the size of such mission, it cannot adequately provide such protection. In that case it shall make available to the head of the mission the information in its possession concerning the location of minefields, mines and booby-traps in that area.

Article 9 — *International co-operation in the removal of minefields, mines and booby-traps*

After the cessation of active hostilities, the parties shall endeavour to reach agreement, both among themselves and, where appropriate, with other States and with international organizations, on the provision of information and technical and material assistance — including, in appropriate circumstances, joint operations — necessary to remove or otherwise render ineffective minefields, mines and booby-traps placed in position during the conflict.

[Technical annex omitted]

* * *

APPENDIX B

CAMBODIAN REFUGEE CAMPS IN THAILAND

The returning Cambodians will come from sven U.N.-assisted camps just inside the Thai border: Site 2, Site B, Site 8, and Site K. For the past eight years, six of these camps, have been assisted by the United Nations Border Relief Operation (UNBRO), an agency created to work specifically with displaced Cambodians in Thailand, as well as a large number of nongovernmental organizations.[104] Each camp also maintains its own Khmer administration. Most of these refugees lead lives of stupefying boredom, bound to whichever political faction happens to administer them and allowed no choice as to affiliation. Because the Thai government has classified them as "displaced persons" or "illegal immigrants" rather than as refugees, they are denied the protection accorded refugees by international law.

Besides the border camps, a UNHCR-controlled camp north of Aranyaprathet called Khao-I-Dang holds about 11,600 Cambodians whose status hangs in limbo. The camp was originally a holding center for officially-accepted Cambodian refugees awaiting resettlement in other countries. But increasingly, it became a camp made up of persons who had been rejected for resettlement. In 1988, the Thai government declared that the remaining camp residents would be eventually transferred to the border for eventual repatriation.

Site 2, with over 200,000 inhabitants, is by far the largest border camp.[105] It is a bamboo city controlled by the KPNLF, whose soldiers, numbering between 6,000 and 8,000, are quartered

---

[104]By early 1980, a total of 95 nongovernmental organizations were working at the border. Of these, 37 were working in the UNHCR-protected camp at Khao I Dang.

[105]Sok Sann, a smaller KPNLF civilian camp, is located along the southern border and mainly functions as a for its military satellite.

at a nearby military base. Because it is located less than one kilometer from the border, Site 2 has frequently been shelled by Cambodian forces. Crime and domestic violence are so widespread in the camp that an UNBRO official has described it as a "time-bomb on a short fuse." Weapons, including handgrenades, are easy for camp residents to obtain. Residents, for instance, told our delegation in April 1991 that a handgrenade could be purchased for Thai Baht 30 (US$ 2). On June 12, 1990, two children died when a grenade was tossed into a video viewing room. The month before our visit, several men threw grenades into a crowd at a dance near the camp pagoda, killing 20 people.

Site B, also known as Green Hill, is a FUNCINPEC camp of over 60,000 located near the summit of the Dangrek escarpment. In early 1990, FUNCINPEC moved its military base into a so-called "liberated zone" inside Cambodia. The camp has fewer of the security and social problems of Site 2, partly as a result of an oppressive administration. FUNCINPEC maintains a fighting force of about 6,000 soldiers who mainly operate in Oddar Meanchey and Battambang provinces.

There are three main "clusters" of Khmer Rouge camps along the border: O Trao, Site 8, and Site K. The Khmer Rouge also maintains camps near Pailin and Phnom Malai inside Cambodia.[106] Except for the last, each cluster typically contains one "open" UNBRO-assisted camp on the Thai side of the border; one or more "closed" camps housing porters and families of fighters hidden close to the border; and several secret military camps inside Cambodia.

O Trao, sometimes referred to as the North Sisophon battlefield, is the northern-most camp along the border and consists of an UNBRO-assisted camp and at least seven satellite camps. Ta Mok, the infamous Khmer Rouge officer, commands over 9,000 Khmer Rouge fighters who operate primarily in the provinces of Oddar Meanchay, Siem Reap, and Preah Vihear. Ta

---

[106] See "Violations of the Laws of War by the Khmer Rouge," *News from Asia Watch,* April 1990.

Mok is reputed to have lost a foot after stepping on a land mine laid by one of his own soldiers.

Site 8, located south of Aranyaprathet, is the largest of the Khmer Rouge camps, with a population of around 38,000. There are three civilian satellite camps, bringing the total civilian population within the Site 8 system to about 55,000. Three other satellite camps are military bases with a total troop strength of 7,000 to 8,000. Like Site 2, Site 8 has been subject to frequent shelling and, in the summer of 1989, had to be evacuated several times because of the intensity of the shelling. Until recently, the camp was restricted to visitors, but over the past two years it has been opened up as a "showcase" by its Khmer Rouge administration. Troops quartered in the Site 8 system operate primarily in Battambang, Siem Reap, and Kompong Thom provinces.

Another Khmer Rouge camp, Site K, is much less accessible to the international community, although UNBRO, ICRC, and some relief organizations have access. The self-imposed isolation of these camps has meant that the inhabitants are denied regular health services, such as basic vaccinations against common childhood diseases, sometimes with devastating results.[107]

The Site K camp system, also referred to as the Koh Kong or Southwest battlefield, contains several secret military camps. Pol Pot lives near the camp in an area called Zone 87 (one of Pol Pot's aliases is "87"). According to former Khmer Rouge commanders who have attended Zone 87 training sessions, Pol Pot remains the unquestioned political and military leader of the Khmer Rouge, formulating policies with the same inner circle that ruled Cambodia in the mid-1970s. The Site K system contains about 25,000 civilians and between 8,000 and 9,000 soldiers who operate in Koh Kong, Kompong Speu, Kompong Chhnang, and Kompot provinces.

---

[107]See John R. Rogge, "Return to Cambodia," p. 59.

In recent years, as the three resistance groups have gained control of more territory inside Cambodia, they have forcibly moved people from the camps into so-called "occupied" or "liberated zones." In 1990, the Khmer Rouge alone moved at least 100,000 people into its own harsh "liberated zones" inside Cambodia.

On several occasions, international relief organizations on the border have issued statements or sent letters to the various factions protesting these forced repatriations to Cambodia. For instance, in November 1990, the ICRC proclaimed that "Under no circumstances should the border population be encouraged or pushed to move into areas where people's health and safety are clearly put at risk...."[108] Some relief workers have also criticized aid organizations that provide relief to refugees who have been forced by the various political factions back into Cambodia. Says Susan Walker, Handicap International's director in Bangkok, "If these nongovernment organizations go into the [liberated] zones, they will only draw more people in and expose them to security dangers, mines, and malaria."[109]

---

[108]ICRC statement on forced repatriation, November 1990.

[109]Interview, Bangkok, April 29,1991.

APPENDIX C

The following is the questionnaire for the mines survey conducted in Afghanistan by the Mines Advisory Group.

MINES ADVISORY GROUP

FIELD QUESTIONAIRE

01. AGENCY DETAILS

011. Agency Name

012. Contact Person

013. Telephone No

014. Indicate main fields of work involving your agency in the subject area : (enter number in box)

[ ] [ ] [ ]

1.Medical 2.Reconstruction 3.Education

4.Agriculture 5.Road Repair/Upgrading

6.Canal/Karez Repair/Cleaning 7.Veterinary

8.Commodity Inputs 9.Emergency Relief

015. Specify other activities not included above!

016. How long has the agency been operational in the subject province?

YEARS_______ MONTHS_______

017. Enumerator's Name: [ ]

C/Q1/DC_____

02. LOCATION DETAILS (for completion by field staff)

021.Province:

022. District:

023: Markaz/Target village:

024. Give details of key contacts in the subject area:

(NOTE TO AGENCY DIRECTOR: If you feel that giving details of your contacts will in any way jeopardise your programme disregard 024)

NAME	POSITION	LOCATION

(NOTE: Where contact is a Commander include party affiliation)

024. How many families in subject area?

025. How many <u>farming</u> families in subject area?

026. Classify the present security situation in the subject area!
(Enter number in box) ☐

1. No hostilities for more than six months

2. No hostilities for 3 - 6 months

3. Occasional hostilities

4. Regular hostilities

03. MINES & THE COMMUNITY

031. Do the community consider that mines are a major obstacle in their daily life? YES/NO

032. Do the community find the presence of mines a greater hazard at specific times of the year? [ ] [ ] [ ]

1. Summer
2. When Snow has fallen
3. During Snow Melt
4. Rainy Season
5. Planting Season
6. Harvest
7. Other (specify) [ ]

034. When a member of the community finds a mine what action is normally taken? [ ] [ ]

1. No action taken
2. Tell family and neighbours
3. Report location to Mujahideen
4. Report to shura
5. Mark with stones
6. Mark in another way [ ]
7. Other action [ ]

C/Q1/DC_____

04. MINES - CASUALTIES

041. How many people have been killed or injured by mines in the subject area?

	KILLED	INJURED
(0411) In the last month?		
(0412) In the last year?		
(0413) In the last 2 years?		

042. Of the total killed and injured in the last two years (0413) how many were:

	KILLED	INJURED
(0421) Men age 20 > 40yrs?		
(0422) Men age over 40yrs?		
(0423) Boys age 10 > 20yrs?		
(0424) Women aged 20 > 40yrs?		
(0425) Women over 40yrs?		
(0426) Girls age 10 > 20yrs?		
(0427) Male children under 10yrs?		
(0428) Female children under 10yrs?		

C/Q1/DC_____

C/Q1/DC_____

043. Are there any people who have been injured by mines <u>from the subject area</u> in hospital in Peshawar now? (give details)

NAME	SON OF:	VILLAGE	APROX. DATE OF INJURY	HOSPITAL

044. How many animals have been killed by mines in the last year?

(0441) Oxen/Cattle _____ (0442) Horses ________

(0443) Camels ______ (0444) Mules/Donkeys ________

(0445) Sheep/Goats ________

045. How many vehicles have been destroyed by mines in the last year?

(0451) Trucks _________

(0452) Tractors _________

(0453)Others ___________

C/Q1/DC_____

C/Q1/DC_____

05. LOCATION OF MINES

051. Which activities are most seriously affected by the presence of mines? (Enter numbers in boxes)

[ ] [ ] [ ]

1. Work in Fields
2. Canal Cleaning/Repair
3. Karez Cleaning/Repair
4. Travel by Vehicle
5. Travel by Foot
6. Grazing of Animals
7. Collecting Firewood
8. House Repair
9. Other activities [ ]

052. How many farming families are in the area? [ ]

053. How many farming families are affected by mines laid on:

a: Arable Land [ ]

b: Grazing Land [ ]

C/Q1/DC_____

054. How many canals are mined? [ ]

055. How many karezes are out of use;

(0551) Because the karez itself is mined? [ ]

(0552) Because the karez cannot be reached for cleaning/repair due to mines? [ ]

056. How many villages;

(0561) Have mines in or near houses? [ ]

(0562) Cannot be approached because of mines? [ ]

057. Are the mountains and hillsides in the subject district thought or known to be mined?

YES / NO

C/Q1/DC_____

06. TYPES & USE OF MINES & DE-MINING

Target Group - Commanders/Mujahideen

061. What are the most common kinds of anti-personnel mines known to have been laid by Soviet/Regime forces in the district?

062. What kind of mines have the Mujahideen laid in the district? "*"

063. Have any kinds of booby traps been used in the area?

YES / NO / NOT KNOWN

064. Describe any method of mine marking known to have been used in the area by Soviet or Regime forces:

065. Describe any method of mine marking used in the area by the Mujahideen: "*"

*" - THESE QUESTIONS NEED NOT BE INCLUDED IN AREAS WHERE IT MAY BE CONSIDERED A BREECH OF MUJAHIDEEN SECURITY

C/Q1/DC_____

066. Have any attempts been made to de-mine in the area?

YES / NO

067. Who organised these attempts?

07. GENERAL

TARGET GROUP - SHURA & MUJAHIDEEN COMMANDERS

071. Will the Shura and Commanders support a policy of immediate destruction of all mines and devices found?

YES / NO

072. Has anyone in the subject area completed and sent to Peshawar a COMMUNITY MAPPING PACK distributed by Operation Salam?

YES / NO

(0721) Name:

(0722) Aproximate date sent:

Human Rights Watch

Human Rights Watch is composed of Africa Watch, Americas Watch, Asia Watch, Helsinki Watch, Middle East Watch and the Fund For Free Expression

Addresses for Human Rights Watch

485 Fifth Avenue
New York, NY 10017
Tel: (212) 972-8400
Fax: (212) 972-0905

1522 K Street, NW #910
Washington, DC 20005
Tel: (202) 371-6592
Fax: (202) 371-0124

10951 West Pico Blvd., #203
Los Angeles, CA 90064
Tel: (213) 475-3070
Fax: (213) 475-5613

90 Borough High Street
London UK SE1 1LL
Tel: (071) 378-8008
Fax: (071) 378-8029

The board of directors of the Physicians For Human Rights includes H. Jack Geiger, President, City University of New York Medical School; Carola Eisenberg, Vice President, Harvard Medical School; Judy Ann Bigby, Brigham & Women's Hospital; Zenaido Camacho, Baylor College of Medicine; Cesar Chelala, New York City; Richard Claude, University of Maryland; Charles Clements, SatelLife, Boston; John Constable, Harvard Medical School; Robert Cook-Deegan, Institute of Medicine; Felton Earls, Harvard School of Public Health; Paul Epstein, Cambridge Hospital; John Hatch, University of North Carolina; Howard Hu, Harvard Medical School; Maria Judge, Fletcher School, Tufts University; Robert Kirschner, Cook County Medical Examiners Office; Robert Lawrence, The Rockefeller Foundation; Jennifer Leaning, Harvard Community Health Plan; Jonathan Mann, Harvard School of Public Health; Gabriel Otterman, Children's Hospital, Boston; Irwin Redlener, Cornell University Medical Center; Randolph Reinhold, Tufts University School of Medicine; Preston Reynolds, Johns Hopkins Hospital; Stephen Robinson, City University of New York Medical School; Helen Rodriguez-Trias, Brookdale, CA; Linda Rosenstock, University of Washington Medical School; Barry H. Rumack, University of Colorado Medical School; Jane Green Schaller, Tufts University School of Medicine; M. Roy Schwartz, American Medical Association; Jeanne Spurlock, American Psychiatric Association; Kim Thorburn, University of Hawaii School of Medicine; Philippe Villers, Concord, MA; Paul Weiss, Harvard Medical School.

The national advisory committee includes Herbert Abrams, Mary Ellen Avery, John Ball, Thomas Ballantine, Paul Beeson, Robert Butler, Christine Cassel, Linda Clever, Robert Ebert, John Edsall, William Foege, Jerome Frank, Alfred Gellhorn, Larry Gotsin, Lester Grinspoon, Lawrence Hartman, M. Alfred Haynes, Howard Hiatt, David Hodges, Edith Irby Jones, Samuel Katz, Cynthia Krane, Joyce Lashof, Philip Lee, Roy Menninger, Carol Nadelson, Aryeh Neier, Jessica Neuwirth, Gilbert Omenn, Frank Oski, Vivian Pinn-Wiggins, Alvin Poussaint, Julius Richmond, Stanford Roman, Jr., Steven Schroeder, Cecil Sheps, Aaron Shirly, Victor Sidel, Giorgio Solimano, Eric Stover, Mervyn Susser, Harold Visotsky, Bailus Walker, Jr., Levi Watkins.

Physicians For Human Rights
100 Boylston Street
Suite 620
Boston, MA 02116
Tel (617) 695-0041
Fax (617) 623-0307